The Hysteresis Machines

The Hysteresis Machines

DR. S.C. BHARGAVA
(Retired) Specialist (Electromagnetic Phenomena)
Corporate Research and Development Division
Bharat Heavy Electricals Limited [A Govt. of India Undertaking]
HYDERABAD. India

and

Variously Visiting Professor of Electrical Engineering
in
Several Engineering Institutions

CRC Press
Taylor & Francis Group
Boca Raton London New York

CRC Press is an imprint of the
Taylor & Francis Group, an **informa** business

BSP **BS Publications**
A unit of **BSP Books Pvt. Ltd.**
4-4-309/316, Giriraj Lane, Sultan Bazar,
Hyderabad - 500 095

First published 2023
by CRC Press
4 Park Square, Milton Park, Abingdon, Oxon, OX14 4RN

and by CRC Press
6000 Broken Sound Parkway NW, Suite 300, Boca Raton, FL 33487-2742

© 2023 BSP Books Pvt. Ltd

CRC Press is an imprint of Informa UK Limited

The right of S.C. Bhargava to be identified as author of this work has been asserted in accordance with sections 77 and 78 of the Copyright, Designs and Patents Act 1988.

British Library Cataloguing-in-Publication Data
A catalogue record for this book is available from the British Library

ISBN: 9781032406466 (hbk)
ISBN: 9781032406473 (pbk)
ISBN: 9781003354093 (ebk)

DOI: 10.4324/9781003354093

Typeset in Times New Roman
by BSP Books, Hyderabad 500 095

BSP BOOKS

The true logic of this world is the calculus of probabilities

- **James Clerk Maxwell**

Dedicated to

the unforgettable memory

of my parents who

contributed as much towards my research

Contents

Preface

All hysteresis machines of which a variety are commercially available and put to many applications are characterised by the use of a 'hard' or permanent-magnet material in the rotor of the machine in different forms whilst the stator may be similar to that of a 3-phase induction motor or comprise salient poles excited by direct current; for example that in a DC machine. The driving torque in the hysteresis machine is typically related to the loss of power resulting from hysteresis – a highly non-liner, multi-valued magnetic phenomenon - in the active part of the rotor caused by 'time-dependent' mmf in the stator.

Of the various hysteresis machines, the hysteresis motor represents the commonest and most widely used machine. A hysteresis motor which forms the mainstay of a variety of hysteresis machines is characterised by *two* important features

(a) it is a *self-starting* synchronous motor in which the torque remains constant with motor speed;

(b) the operation of the machine is quiet owing to absence of any slots on the rotor and a winding housed therein – an essential requirement for most other electric motors that results in a noisy operation.

The present book, one of its kind, is divided in three parts dealing with various forms of hysteresis machines.

The first part describes the etymology of the phenomenon of hysteresis, its physical interpretation, the basics of the measurement techniques used for plotting hysteresis loops, an idea of various hard magnetic materials and various approximations of hysteresis loops adopted for varied analyses of machines using hysteretic materials.

The second part deals with the design, operation and applications of commonly known and utilised hysteresis machines, viz., hysteresis motor, hysteresis coupling, hysteresis brake, hysteresis clutch and hysteresis-reluctance motor.

The third part is devoted at length to a comprehensive study of an experimental hysteresis machine. The contents of this part derive almost entirely from the experimental and analytical work carried out by the author during his doctoral research at the University of Aston in the UK. The exhaustive range of tests and measurements, obtained by the use of uniquely devised transducers and techniques, many of these having been evolved for the first time, showcase numerous aspects of a hysteresis machine, never before brought out anywhere. These relate mainly to the measured waveforms of radial and peripheral flux density in the *airgap* and active part of the rotor as influenced by the hysteretic properties of the material. This is followed by a detailed description of analyses of production of torque in the machine derived from the measured flux density variation, mainly in the airgap and nonmagnetic region of the machine, involving electromagnetic field theory and use of Poynting theorem.

- The Author

Acknowledgements

The author would like to express his gratitude to Dr.M.J.Jevons, his supervisor for doctoral research at the University of Aston, UK, and an expert par-excellence on all aspects of electromagnetism,

and

Prof. E.J.Davies of the Departmental of Electrical Engineering at The University for his valuable guidance and suggestions, and providing all facilities during the course of research, esp. the assistance of all departmental technical staff for carrying out intricate, exhaustive experimental work.

The author's special thanks are due to his daughter, Anshu, for her untiring help in the preparation of the typescript involving many complicated expressions and mathematical derivations.

Lastly, he is thankful to the publishers of the book for taking good care of the printing and publication.

- The Author

About the Author

Dr S.C. Bhargava graduated in electrical engineering in 1963 and postgraduated in power systems in 1966 from the University of Roorkee, India. He used to teach graduate and (part-time) postgraduate courses at (now) N I T, Jaipur before proceeding to the UK on a University of Aston scholarship for his Ph D which he obtained in 1972. His research was heavily based on analytical and extensive experimental work using electromagnetic fields and their applications. On his return from the UK, he served as Specialist, Electromagnetic Phenomena, at Corp. R & D, B H E L, Hyderabad, from 1974 to 1999 where his research activities involved application of electromagnetic phenomena at large. Following his retirement from BHEL in 1999, he had been a visiting professor at several engineering institutes in Hyderabad, teaching graduate and postgraduate courses.

He is the Fellow or Senior Member of several Professional Societies in India and abroad such as Institution of Engineers (India), Institution of Electrical Engineers (UK), Institution of Electronics and Telecommunication Engineers (India), Institute of Electrical and Electronics Engineers (USA) and Computer Society of India.

His fields of interest include "biological effects of electromagnetic fields on living beings" on which he has published several papers, and "solar electricity".

He has published over 40 technical papers in National and International journals of which many were presented at National and International conferences.

He is the author of the book "Electrical Measuring Instruments and Measurements", published in 2013 by B S Publications in Hyderabad and CRC Press in the UK. The book has been approved by the JNTU as one of the text books on electrical measurements, a subject taught to EEE students of the University in III semester.

His other book, titled "Finite Element Analyses of Eddy Current Effects in Turbogenerators", dealing with special problems of turbogenerators, based on his vast experience related to electromagnetic fields and their application to analyses of turbogenerators, has been recently published by CRC Press, Taylor & Francis Group, UK.

He has recently come out with a unique book titled "A Book of Physics – In Perspective" that deals comprehensively with almost all topics of physics at high school and board levels to help the students and teachers of physics alike to understand the various aspects of physics in *perspective*.

Symbols and Notation

A	Leading pole tip of the field system
B	Lagging pole tip of the field system
B	Flux density*
$\hat{B}$, B_m, B_{max}	Maximum value of flux density
B_r	Radial flux density component
B_θ	Peripheral or tangential flux density component
ΔB_r	'Difference' flux density output from the full-pitch coils
e	Induced electromagnetic force (EMF)
E	Electric field intensity
E_z	Axial component of electric field intensity
F, f	Force
H	Magnetising force or field
$\hat{H}$, H_m, H_{max}	Maximum value of magnetising force
H_r	Radial component of magnetising force
H_θ	Tangential component of magnetising force
I	Excitation current
J	Magnetic current density
l	Axial length of the machine
M_A	Applied magnetomotive force (mmf)
M_R	Rotor mmf
M	Rotor magnetisation vector
$\mathcal{M}$	Magnetic dipole moment
N	Speed
n	Order of harmonic

---

*Vector quantities are denoted by a (bar) above the appropriate letter as $\bar{B}$, or associated with unit vectors as $\bar{Bi}_n$, $\bar{Bi}_r$, $\bar{Bi}_\theta$, $\bar{Bi}_z$ etc.

P	Power
R, r	Radius
r	General suffix in radial direction
S	Poynting vector
	Poynting vector surface
S_r	Poynting vector in radial direction
S_θ	Poynting vector in peripheral direction
T	Torque
T_s	Maxwell tensile stress
W_h, W_{alt}	Alternating hysteresis loss
W_{rot}	Rotational hysteresis loss
z	General suffix in the axial direction
α	Angle of incidence of resultant flux density on rotor surface
α'	Angle of refraction of resultant flux density on rotor surface
α', α'', β', β''	Magnetic surface current elements
β	Angular displacement
Δ	Incremental value
δ	'Load' angle of the machine
θ	Angular distance
	General suffix in the peripheral direction
μ_r	Relative permeability
λ	Flux linkage
ϕ	Magnetic scalar potential
$\varnothing$	Magnetic flux
ρ	Resistivity
ψ, Ψ	Conjugate scalar magnetic potential
ω	Angular speed of rotation
τ	Torque

$$\overline{\nabla} = \overline{i}_r \frac{\partial}{\partial r} + \overline{i}_\theta \frac{1}{r} \frac{\partial}{\partial \theta} + \overline{i}_z \frac{\partial}{\partial z}$$

Part A
INTRODUCTION

I : Introduction

1

Introduction

ETYMOLOGY AND HISTORY

The Word Hysteresis

The word "hysteresis" is of Greek origin, spelled as $\nu \; \sigma \; \tau \; \varepsilon \; \rho \; \varepsilon \; \omega$ in Greek alphabet, meaning to "lag behind" . The dictionary meaning of the word is given as n. phenomenon whereby changes in an effect lag behind changes in its cause. Or a retardation of the effect when the forces acting upon a body are changed (as if from viscosity or internal friction); *esp.* lagging in the values of resulting magnetisation in a magnetic material (as iron) due to a changing magnetising force. -**hys-ter-et-ic** *adj.*

Whilst commonly used in relation to magnetic effects in ferromagnetic materials, hysteresis as defined above can be found in physics, chemistry, engineering, biology and economics and so on; for example, in the deformation of rubber bands and shape-memory alloys and many other natural phenomena. In natural systems it is often associated with irreversible thermodynamic change such as phase transitions and with internal friction; dissipation is a common side effect.

Hysteresis can be a *dynamic lag* between an input and an output, known as *rate-dependent* hysteresis; phenomena such as the magnetic hysteresis loops are mainly rate-independent.

Hysteresis in Magnetic Materials

The phenomenon of hysteresis was first associated with magnetic materials by Warburg[1]*, and simultaneously by (Sir James Alfred) Ewing[2], about the year 1879. Until then, the use of 'iron' – the commonest example of magnetic material in those days in electric machines and equipment – was based on the assumption of single-valued relationship between magnetising force and induced magnetism[3] and perhaps the importance of residual magnetism was not fully realised. Ewing showed that the induced magnetism lagged behind the applied magnetomotive force (mmf) and named this property the "hysteresis effect"[4]. Cyclic variation of the mmf resulted in a closed loop – the hysteresis loop – and once again this was demonstrated on a theoretical basis by Ewing[5] with the help of various models using small permanent magnets pivoted on needle supports.

---

*numbers in square brackets show references, inclusive or otherwise, listed at end.

A typical hysteresis loop pertaining to a 'hard' or permanent magnet, describing its 'key' parts or sections, is shown in Fig.A1.1. In many cases, an inherent accessory of a (major) hysteresis loop is a simple minor loop, called a recoil loop, as drawn in Fig.A1.2[1].

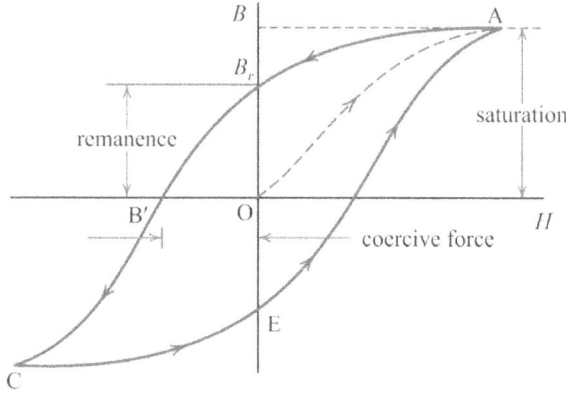

Fig.A1.1 : Hysteresis loop of a permanent magnet

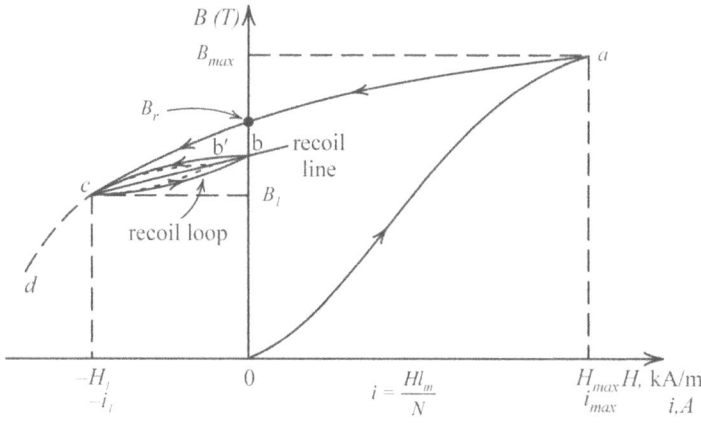

Fig.A1.2 : Demagnetisation curve and a minor loop

Referring to Fig.A1.1, the (initial) part OA represents the "virgin" magnetisation curve of the material till saturation sets in (point A)[2]. At this point if the magnetising force is

---

[1]As shown in Fig.A1.2, a recoil loop identified as "major or "minor" ensues if, say during demagnetisation of a given material such as section B_r c in the figure, the magnetising force is reversed in the 'positive' direction to some value and reversed again to revert to the original B,H point. This is discussed further in Appendix I.

[2]It is well-known that the curve is not smooth, but is characterised by increase of magnetisation with applied magnetising force in 'small', random steps, identified as "Barkhausen jumps". Discovered by German physicist Heinrich Barkhausen in 1919, it is caused by rapid changes of size of magnetic domains (similarly magnetically oriented atoms in ferromagnetic materials).

removed, the magnet is left magnetised with a *remanence* or "residual magnetism", the intercept OB_r, theoretically sustained indefinitely. If the magnet were to be demagnetised to reduce the remanence to zero, the magnetising force must be reversed, section OB', being defined as coercive force (or "coercivity" of the material). A further increase of the magnetising force in the reverse direction would result in magnetisation of the magnet with reverse polarity, up to saturation, shown by point C. If so desired, the process of demagnetisation and magnetisation to original condition would trace the path depicted by C E A. The complete cycle is identified as a hysteresis loop and is *material* dependent[3]. This is illustrated qualitatively in Fig.A1.3 for three samples of magnetic materials.

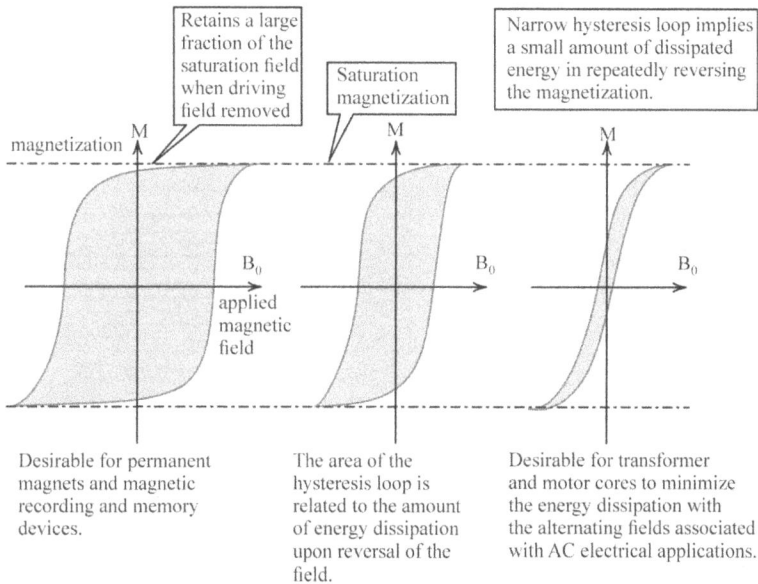

Retains a large fraction of the saturation field when driving field removed

Saturation magnetization

Narrow hysteresis loop implies a small amount of dissipated energy in repeatedly reversing the magnetization.

magnetization M

M

M

B_0

B_0

B_0

applied magnetic field

Desirable for permanent magnets and magnetic recording and memory devices.

The area of the hysteresis loop is related to the amount of energy dissipation upon reversal of the field.

Desirable for transformer and motor cores to minimize the energy dissipation with the alternating fields associated with AC electrical applications.

Fig.A1.3 : Hysteresis loops for 'hard' and 'soft' magnetic materials

PHYSICAL INTERPRETATION

Hysteresis loss

The physical meaning and importance of the hysteresis loop was again explained by Warburg[1]. It was established by further research[6],[7] that the *area* of the hysteresis loop represented the *energy* 'lost' as heat in taking a material through one complete cycle of magnetisation[4]. The energy is expressed in joules.

---

[3]Clearly, 'soft' magnetic materials such as sheet steel that are commonly employed in electric machines would exhibit only feeble hysteretic property, or a very 'narrow' hysteresis loop.

[4]Analytically, the total loss for a magnet of volume V during one complete cycle (or one hysteresis loop) would be

$$\text{loss} = V \times \int H \, dB$$

the integral representing the area of the loop.

Alternatively, a permanent magnet will contain magnetic energy for any application, represented by the 'size' of hysteresis loop of the magnet material: the 'fatter' the loop, larger will be the energy content. In this respect, the hysteresis loop is akin to the "indicator diagram" of an internal combustion engine. It follows that the sub-area of the loop that would represent 'strength' of a magnet once magnetised and put to any use would be the area $B_r B' O$, Fig.A1.1; the larger the intercept OB, the remanence, and larger the coercivity, the stronger will be the magnet.

Steinmetz's Expression

Steinmetz[5] suggested a frequently used empirical expression, based on his extensive laboratory experiments, to estimate hysteresis loss of a material that yields satisfactory results in most applications. The expression is given by

$$W_{hyst} = \eta \ B_{max}^{1.6} \times f$$

per unit volume of the material and where B_{max} is the maximum flux density pertaining to a given hysteresis loop; for example, denoted by Part A in Fig.A1.1 and f denotes the cycle of magnetisation. η is called "Steinmetz coefficient", being material dependent.

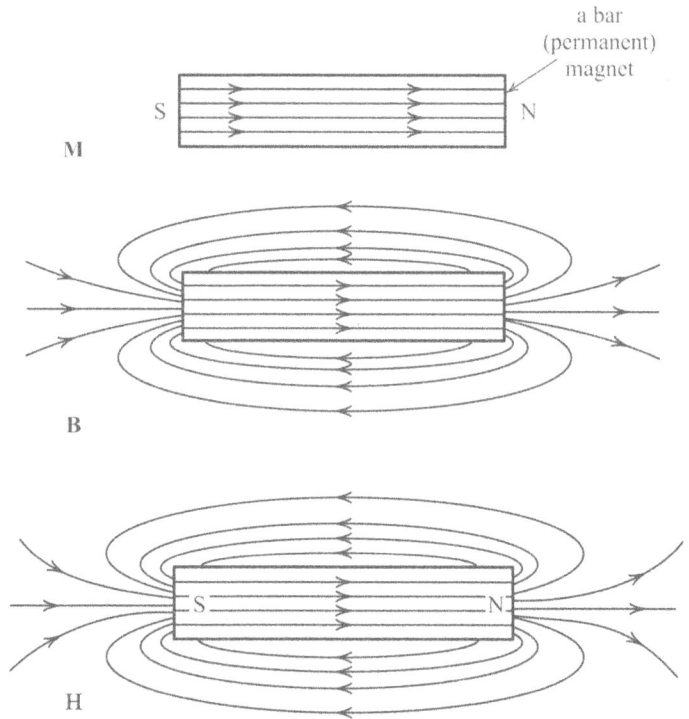

Fig.A1.4 : The directional properties of magnetic flux density and magnetisation in a permanent magnet

In this context, it is important to observe how the intensity of magnetisation, M, the magnetising force, H, and resultant magnetic flux or flux density, B, are directed with respect to each other. This is shown qualitatively for a bar magnet in Fig.A1.4.

It is seen that the flux lines are *oppositely* directed to the direction of magnetising field within the magnet.

---

[5]Charles Proteus Steinmetz (1865-1923).

Experimental Determination of Hysteresis Loops

It often becomes necessary to obtain *actual* hysteresis loops, major as well as minor[6], of a permanent magnet material by performing suitable experiment before its in an application. This is because the loops provided by the manufacturer may be somewhat at variance with the actual hysteresis loops. Further, many materials may be resorted to some degree of machining etc; for example, flexible or those in strip form being formed to, say, an annulus and may develop slightly different magnetic properties, esp. following a recommended heat treatment.

Experiments on permanent magnet materials to determine their magnetic properties, esp. obtaining hysteresis loops, are not simple and straightforward and may often require rather tedious methods or procedures.

Use of hysteresigraphs

A commonly employed device to obtain hysteresis loops of a permanent magnet material using an appropriate sample is known as hysteresisgraph. A typical hysteresigraph, shown schematically in Fig.A1.5, consists of

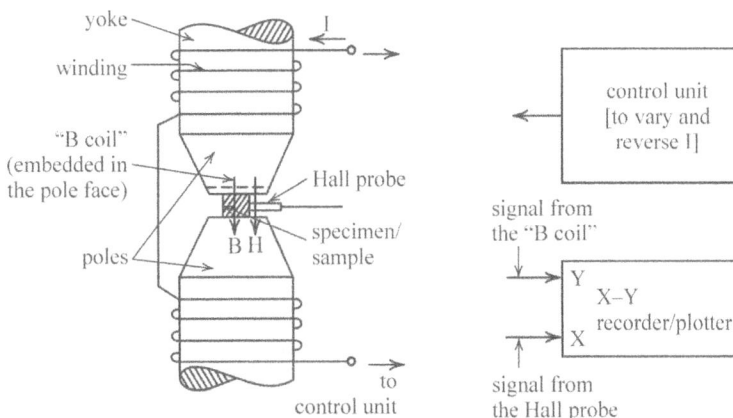

Fig.A1.5 : Schematic of a hysteresigraph

(i)   A "massive" electromagnet, each of its poles excited by direct current passing through a winding of large number of turns that can carry sufficiently high current; the poles of the electromagnet being usually of (truncated) conical shape, about 8 cm in diameter at ends. A built-in mechanism in the device allows the

---

[6]As discussed, a minor loop ensues if during demagnetisation of a magnet, the magnetising force is reversed - inadvertently or deliberately - in the 'positive' direction to 'some' value and reversed again to revert to the original B,H point. Referring to Fig.A1.2, the recoil loop b c is more identified as a "major" recoil loop, the tip b reaching (or in constant with) the "B" axis a minor recoil loop may be identified as c b′ , shown dotted.

poles to be moved up and down in a vertical direction to vary the airgap between them, and lock in that position.

(ii)   A control mechanism to vary the excitation current from zero to a very large value to produce a magnetising field of, say, 300 to 500 kA/m so as to drive the sample material well into saturation, or demagnetise it if required, and to reverse the direction of current to apply magnetising field in the reverse direction.

(iii)   Devices or means to measure magnetising field and corresponding flux density in the sample. The former is usually measured using a calibrated Hall probe, positioned appropriately in the vicinity of the sample within the airgap whilst the latter is invariably measured with the help of specially designed "B coils" (or search coils), embedded in one of the poles (usually the lower pole), or by winding a search coil around the sample.

(iv)   An X-Y plotter or recorder to which are fed the outputs from the Hall probe and "B coil", respectively, to the X and Y terminal pairs to plot the B-H curve followed by the hysteresis loops[7].

The details of an *actual* hysteresigraph and results obtained from a sample of permanent magnet material and other means of plotting hysteresis loops (using a permagraph) are described in Appendix I.

Alternating and Rotational Hysteresis

In practice, two types of hysteresis are known to occur in magnetic materials: alternating and rotational. The foregoing discussion applies to alternating hysteresis in which the cycle of magnetisation consists of increasing the magnetising force in one direction to a maximum value, usually to result in saturation of the material, reducing it to zero, followed by a similar variation in opposite direction, as shown in Fig.A1.1. In contrast, rotational hysteresis results from the rotation of a magnetic material, say a disc, through one cycle in an applied magnetising force of *constant* magnitude. This was predicted by Swinburn in 1890[8] and confirmed by Baily in 1894[9]. In most discussions and applications, it is the alternating hysteresis that is more relevant and takes precedence.

PERMANENT MAGNETS

Modern permanent magnet devices require the use of materials possessing large coercive forces; in turn this usually requires the presence of magneto-crystalline anisotropies. This is achieved these days by the use of rare-earth-based materials that possess sufficiently large anisotropies where this property originates from a combination of the crystal-field interaction of the 4f electrons with electrostatic charge of the surrounding ions and the relatively strong spin-orbit interaction of the 4f electrons. More than a decade ago high-performance permanent magnets were based on the use of alloys containing samarian (Sm) and cobalt (Co); such as $SmCo_5$. Recently, an even more powerful permanent

---

[7]The modern means to record the B-H curve and hysteresis loops may entail use of A-D converters and digital recording.

magnet material has been discovered which is based primarily on the ternary intermetallic compound $Nd_2Fe_{14}B$.

From the time of their historic developments decades ago, some of the commonly employed materials and their characteristics are summarised below.

Table A1.1: Permanent magnet materials

Material	Composition	Characteristics
Ceramic, also known as ferrite	iron oxide and barium or strontium carbonate	energy up to 23,800 J/m^3 brittle, low cost, high coercive force, high resistance to corrosion
Alnico V	aluminium, nickel and cobalt	energy up to 37,400 J/m^3 high cost, high corrosion resistance high mechanical strength, high temperature stability, low coercive force, good resistance to demagnetisation
Samarium Cobalt	samarium and cobalt (rare earth magnetic material)	energy up to 102,000-136,000 J/m^3 can work up to 300°C high cost, high corrosion resistance, high temperature stability, high coercive force, low mechanical strength-brittle
Neodymium-iron-boron	neodymium, iron and Boron (another rare earth material)	energy up to 340,000 J/m^3 much higher cost, high coercive magnetic force, low mechanical strength-brittle, moderate temperature stability, low corrosion resistance
Injection moulded	resin and magnetic material powder	moderate energy product high cost, low temperature stability, moderate coercive force, high corrosion resistance highlyshapeable – can be manufactured in complex shapes owing to process of injection moulding
Flexible	same as injection moulded	available in strip and sheet form low energy product, low cost, high corrosion resistance, low stability, moderate coercive force

An alloy introduced in 1970s offering many special features is known as "vicalloy".

Vicalloy[8]

Vicalloy is one of the family of cobalt-iron-vanadium based permanent magnet alloys. The alloy primarily consists of vanadium (10%), iron (about 37.4%) and cobalt (52%); the rest being manganese, carbon, silicone and a few more elements in very small quantities. The alloy is available commercially in the form of

- round bars of diameter 0.1 mm (wires) to 100 mm and random length
- strips or sheet of thickness 0.1 to 2.5 mm and width from 5.5 to 180 mm

---

[8]Used exclusively and extensively in the experimental hysteresis machine, discussed later in Part C.

Its development in commercially usable form is due to Nesbitt[10][9] who also held the original patent.

The material in strip form can be machined or bent in any shape; for example, an annulus, *before heat treatment*. Following the recommended heat treatment[10], the material becomes "glass hard" and cannot be machined or formed further.

Vicalloy has been used variously in hysteresis machines[11].

Magnetic properties of vicalloy
- residual magnetism, B_r: 0.9 T
- coercivity, H_c: 0.3×10^5 A/m
- energy product: 8,000 J/m^3
- resistivity: 0.75 $\mu\Omega$-m[12]

Full magnetic characteristics of vicalloy are considered in Appendix I.

Applications of Permanent Magnets
Permanent magnets find extensive applications typically in*
- DC motors, esp. permanent magnet type
- Synchronous motors
- Stepper motors
- A variety of hysteresis machines
- Moving coil actuators
- Holding force actuators
- Magnetic suspensions
- Sensors
- 'Steady field' providers

---

*The list is suggestive, not exhaustive.

---

[9]See, for example, E. A. Nesbitt: Vicalloy – a workable alloy for permanent magnets, Trans AIMME, Vol.166, 1946, pp 416-25.

[10]Heating to 600°C ± 2°C in an inert or vacuum oven, followed by gradual cooling to room temperature.

[11]For example, vicalloy has been in use in hysteresis motors both in solid and laminated form (for higher frequency applications).

[12]Although the energy product of vicalloy is low in comparison to, say, ALNICO V and far lower than those of rare earth materials, it was chosen for hysteresis machine (Part C) on account of its availability in strip for, ease of machining (including drilling of holes for search coils etc.) prior to heat treatment and high resistivity to minimise eddy-current loss in the annulus.

Hysteresis Loop Approximations

The implicit property of any magnetic material, esp. permanent magnets, results in highly non-linear relationship in magnetising force 'H', and corresponding induced flux density, B, more significantly in saturated condition of the material. The relationship cannot be expressed by a single mathematical equation or form. This assumes a more complicated form when considering closed hysteresis *loops,* following multi-valued relationship between H and B, esp. when required to be modelled in analytical treatment of operation and performance of machines and devices incorporating permanent magnet materials. This has led to various simplifications or approximations of hysteresis loops whilst modelling various machines. Some of these are qualitatively illustrated in Fig.A1.6.

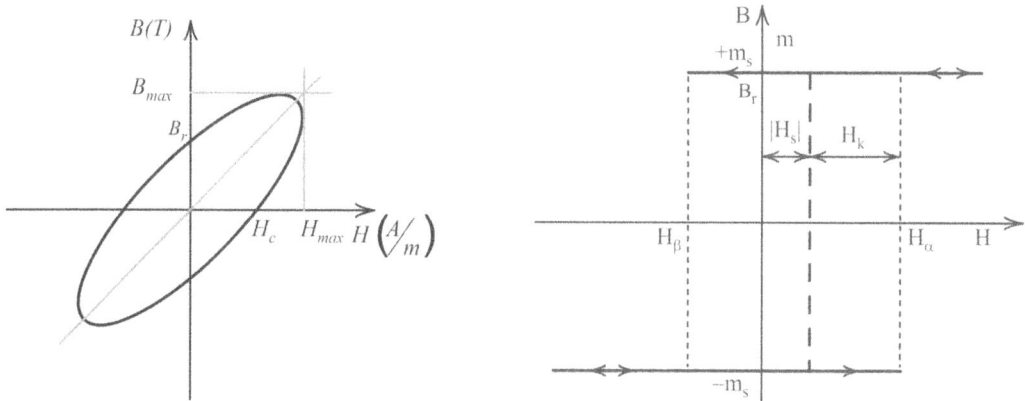

(a) hysteresis loop as an inclined ellipse   (b) rectangular approximation

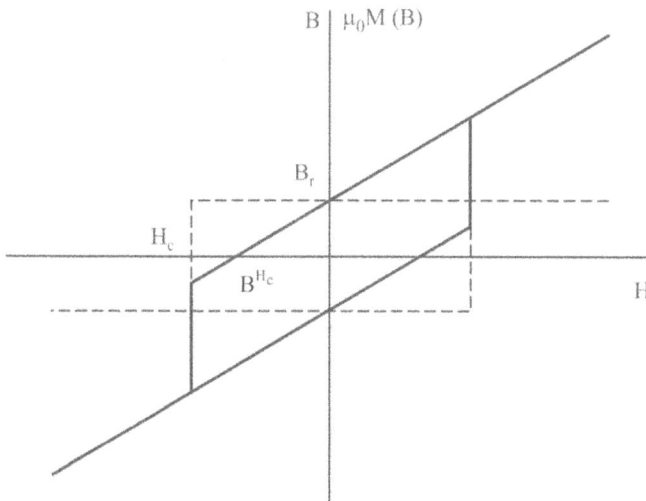

(c) Hysteresis loop as a parallelogram

Fig.A1.6 : Hysteresis loop approximations

An approximation that has been in use to effect some distinct advantages is the "inclined ellipse".

Inclined ellipse

This approximation has been widely used [11], [12], [13], [14], [15] corresponding to an *assumed sinusoidal variation of flux density* and considering only *fundamental* component of applied mmf. The elliptic hysteresis loop, inclined to the mmf axis, permits inclusion of the angle of hysteretic advance consistently in the analysis based on the solution of Laplace's equation[15], resulting in a relatively simple analysis. Whilst the elliptic hysteresis loop approximation may retain nearly the entire area of the actual hysteresis loop, and hence the energy product, and may be easy to model analytically, a source of inaccuracy may arise from the assumption of a similar variation for higher-order space harmonics in the applied mmf and the subsequent superposition of, say power loss, in the analysis. Since the harmonics rotate at different speeds relative to the fundamental, these do not necessarily form closed loops of elliptical shape and the loops may not be repeatable.

The basis of assuming an elliptic hysteresis loop in hysteresis machines to account for *spatial* hysteresis is to express the sinusoidal "B" wave being shifted in space by a *constant* angle 'γ'. For any point of the rotor, there are thus two components of H and the cycle of magnetisation is elliptic. Thus,

$$\begin{bmatrix} B_r \\ B_\theta \end{bmatrix} = \mu \times \begin{bmatrix} \cos\gamma & \sin\gamma \\ -\sin\gamma & \cos\gamma \end{bmatrix} \begin{bmatrix} H_r \\ H_\theta \end{bmatrix}$$

with
$$B = \sqrt{B_r^2 + B_0^2} = \mu\sqrt{(H_r^2 + H_\theta^2)}$$

The assumption of constant γ and μ allows the hysteresis to be accounted for, and an analysis using Laplace's equation would be applicable.

Parallelogram approximation

This was first used by Copeland and Slemon [16], [17], later modified to a rectangular shape to represent the steeply rising part and saturation stage of the B-H curve. Beyond the 'knee' of the curve, the saturation condition of the material is simply regarded as the "airgap line" with unity relative permeability. The advantages claimed of this approximation were analytic simplification and the feasibility of deriving an equivalent circuit for the machine using the particular material. However, an obvious limitation would be that the flux distribution would contain large number of harmonics even when the applied mmf would be assumed to be sinusoidal.

Note that both approximation shown in Fig.A1.4(b) and (c) pertain to the above discussion.

Other Approximations

At low magnetising forces, hysteresis loops formed by displaced parabolas may justify close resemblance to the actual loops and provide an accurate means to estimate alternating hysteresis loss analytically[18]. A displaced rectangular hyperbola shape may seem to be more appropriate approximation for loops having B_m, H_m point near 'knee' of the B-H curve.

Machines Incorporating Hysteretic Materials

There are a variety of machines incorporating permanent magnetic material, almost invariably being a part of the rotor. The characteristic aspect of these machines is to exploit the hysteresis loss in the material to result in useful power or torque in the machine; in some cases the torque may be a braking torque.

Some of these machines such as

- the Hysteresis Motor
- the Hysteresis Coupling
- the Hysteresis Brake
- the Hysteresis Clutch
- the Hysteresis-reluctance Motor

are described later in Part B.

Part B

HYSTERESIS MACHINES: GENERAL

1 : The Hysteresis Motor

1

The Hysteresis Motor

Basics

The two distinct features that characterise a hysteresis motor as compared to other AC motors, esp. the most commonly used induction motor are

A. a hysteresis motor is a self-starting synchronous motor; its speed being governed by the supply frequency and number of poles for which the three-phase winding is designed.

B. the torque developed in the motor is theoretically constant from the instant of start to attaining the synchronous speed, locked to frequency of supply.

History

Although no clear evidence is available, the origin of 'hysteresis motor' can be traced back to 1886-87. In one of the well-known researches described in "The electrodynamic rotation produced by alternating currents"[19,20], Ferraris used a laminated iron cylinder and a rotating magnetic field leading to rotation of the cylinder in the direction of the rotating field. The torque that was developed in the cylinder could not be explained completely due to eddy currents and was stated by Ferraris to result principally from hysteresis effect. This elementary device which he called a "motor" can thus be considered as the original hysteresis motor. According to Ferraris, such a motor would be of little industrial use because of the low value of the developed torque and yet might find some useful applications on account of its special properties, viz., self-starting synchronous motor with a sub-synchronous torque which, ideally, is independent of speed.

About the time of 'accidental' invention of hysteresis motor was the development of an instrument by Ewing[12] of historical importance. The device, called the "Hysteresis tester", was devised to measure the hysteretic property of a laminated magnetic material in terms of angle of hysteretic advance. Torque was produced by the action of the field of an electromagnet on a mechanically rotated sample and was clearly due to hysteresis effect in the absence of eddy current[1].

---

[1]On the basis of torque production, Ewing's hysteresis tester can be regarded as another historic form of a hysteresis 'machine' and, if the electromagnet is allowed to rotate, as the first hysteresis coupling.

The schematic cross-section of a 'small' (about 7 cm long) hysteresis motor from 'early' times depicting the arrangement of stator winding and other parts is shown in Fig.B1.1(a); a photo of a 'small' commercial motor, about 7 cm long, is given in Fig.B1.1(b). The motor developed a torque of 500 gm-cm at 3,750 rpm[2].

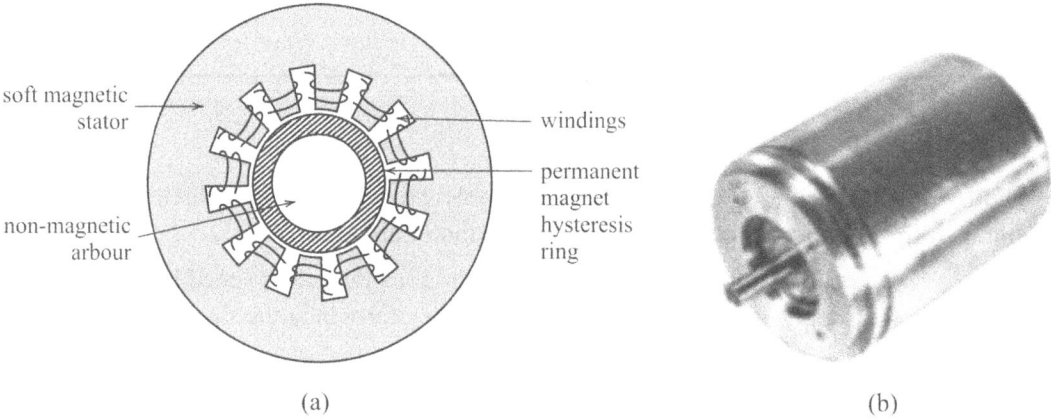

soft magnetic stator

non-magnetic arbour

windings

permanent magnet hysteresis ring

(a)   (b)

Fig.B1.1 : A small hysteresis motor from an early design

STEINMETZ'S THEORY

A theory of the hysteresis motor was propounded by Steinmetz circa 1896-97[21,22]. He derived an expression for the developed torque and efficiency of the motor based on alternating hysteresis, the machine being treated as an ideal model.

The principal features of his theory can be summarised as follows:

(a) The developed torque is proportional to hysteresis loss in the rotor material and hence to the area of the hysteresis loop[3]

(b) Assuming a sinusoidal mmf distribution, the efficiency is proportional to the "angle of hysteretic advance"

(c) During asynchronous operation, the electric power supplied to the rotor divides in two parts: a part, proportional to slip, supplies hysteresis loss that is dissipated as

---

[2]See, for example,

D.R. Driver: Magnetic alloys for hysteresis motors, Publication R16, Telcon metals Ltd., reprinted from ELECTRICAL TIMES, 10 August, 1967.

[3]Note, however, that whilst the area of the hysteresis loop, and hence the loss per cycle, is dependent on the magnitudes of remanence (or residual flux density) and coercivity of the material, a desirable combination of these parameters would be to choose a material with high value of remanence and small coecivity, *for the same loss per cycle*, necessitating low source mmf. Vicalloy is such a material. By the same reasoning, a ferrite magnetic material is hardly suited for a hysteresis machine, apart from the difficulty of machining or forming.

heat; the remainder, proportional to speed supplies the mechanical power at the shaft[4].

PROGRESS DURING THE FIRST-HALF CENTURY
Developments Following Steinmetz's Work

Hysteresis motors attracted very little attention during the early years of the twentieth century since their 'practical' use was much limited in comparison to other contemporary motors; for example, the DC motor and the usual synchronous motor. Nevertheless, the effect of hysteresis in rotors of polyphase induction motors, in modifying their characteristics or affecting their performance, was widely reported[23, 24, 25]. It was observed that the operation of an induction motor near synchronous speed or "passing through synchronism" was noticeably influenced by the hysteresis torque, probably being comparable to that due to normal induction-motor action, esp. at very low values of slip[5].

The fact that the hysteresis in the rotor of an induction motor could give rise to useful torque was first noted by Smith[26] and then, specifically, reported by Robertson in 1911[27] who stated that if only the torque due to hysteresis could be made large enough, the motor would run without slip, that is at synchronous speed[6]. There would still be the undesirable hysteresis loss due to high-frequency pulsations in the teeth, but this could be minimised by using closed or partially-closed rotor slots.

The possibility of a 'pure' hysteresis motor on a commercial scale was discussed by Robertson on the basis of high-hysteresis loss materials then available, but was proved to

---

[4]Thus, if P_h is the hysteresis loss per unit volume given by $P_h = k_h f_r B_m^n$ or $P_h = k_h s \, f_s B_m^n$ where
k_h is the hysteretic coefficient depending on the rotor material
f_r frequency in rotor reference frame
f_s supply frequency
s the slip
B_m the (max) rotor flux density and
n an integer
Then the hysteresis torque will be given by $T_h = P_h/s \, \omega_s$ where ω_s is 'synchronous' angular speed
Substituting,
$T_h = k_h B_m^n / 2\pi$ where $\omega_s = 2 \pi f_s$
And is independent of supply frequency or speed of the rotor.

[5]The magnetic material(s) employed for the construction of rotor used to be of 'poor' quality with appreciable hysteresis loss.

[6]In theory providing at least the needed torque to supply mechanical losses such as friction and windage even if not providing much useful torque.

be impracticable[7]. The power factor in such a motor would be invariably low and input current relatively high, resulting in poor efficiency[8].

TEARE'S THEORETICAL APPROACH

Teare's paper in 1940[29] was a fresh start to develop a detailed theory of hysteresis motor torque, based on the principle of "virtual work"[9]. A general expression for the torque was derived that accounted for rotor flux density in all three *space* directions. More specifically, a simplified torque equation for a 'thin' rotor was obtained which supported Steinmetz's theory; this assumed that the radial component of flux density would not contribute noticeably to the developed torque. The torque expression was verified experimentally by comparing test results from a "circumferential-flux" hysteresis motor[10].

A valuable part of Teare's theory is the treatment of motor operation at synchronous speed with particular reference to varying load conditions and the effect of odd space harmonics. The latter are shown to modify the fundamental hysteresis loop, giving rise to wave like indentations superposed on the main loop, resulting in a reduction of the developed torque as illustrated in Fig.B1.2(a). Changes in load requirements at the shaft at synchronous speed are explained by 'adjusting' the *width* of hysteresis loop to change its area as shown in Fig.B1.2(b)[11].

---

[7]For example, a motor with one horse power output (746 W) would require from 1.15×10^{-2} to 1.64×10^{-2} m^3 (700 t0 1000 cubic inch) of rotor iron at a flux density of 1 T in the rotor at a supply frequency of 50 Hz.

[8]In spite of their small size, low output and very low efficiency, hysteresis motors found increasing use in several special applications as detailed by Holmes and Grundy[28].

[9]This may be explained in this case as "rate of change of stored magnetic energy with angular displacement" as follows:

the total stored energy in the airgap is $W = \dfrac{\text{volume of airgap} \times \text{average of B}^2 \times 10^7}{8\pi} \text{ Ws}$

then   the torque $= \dfrac{dW}{d\alpha}$, α being the angular rotation

[10]A term increasingly being used from then on to describe a motor with a rotor having non-magnetic arbor. This is discussed in detail later.

[11]This is in accordance with the expression for deriving hysteresis loss using Steinmetz's theory where the torque is given to be proportional to B_m.

(a) effect of space mmf harmonics   (b) change of load at synchronous speed

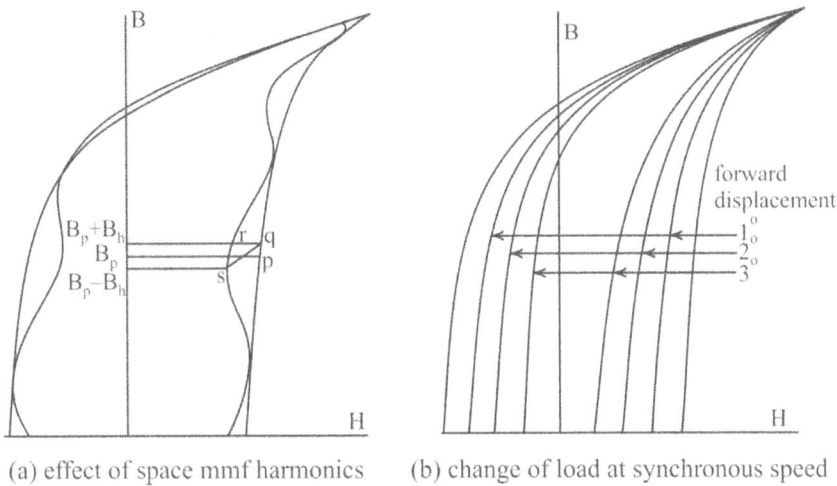

Fig.B1.2 : Hysteresis motor performance

Commercial Improvements

These evolved mainly on account of development of improved 'hard' magnetic materials, mainly alloys containing one or more 'special' elements, having much higher hysteresis loss per unit volume. This kept pace with the efforts to improve performance of the motor by modifying constructional features, notably by 'Smith & Sons'[30] and Rotors[31]. The former adopted a unique design of using magnetic sleeves to bridge the stator slots in order to reduce the undesirable effects of stator-slot-flux undulations in the airgap, the sleeves being cylindrical in shape and inserted into the pre-wound stator.

Modifications on similar lines were reported by Rotors who discussed the effects of open stator slots in detail. The reason of low efficiency of a typical hysteresis motor was attributed by Rotors to the undulations of the space distribution of the rotor flux density relative to the rotating magnetic field resulting in "spurious loss" that could be of the order of one to three times the power output of the rotor. Reduction of this loss was achieved by using a 'large' number of closed slots on the stator, claiming typical efficiencies of 80%; and thus leading to economical manufacture of even fhp motors. To speed up manufacture, Rotors came up with an ingenious scheme of "back winding"[12] of the stator.

However, a machine with closed stator slots would lead to increased magnetising current and low power factor due to the shunting action of the slot bridges. Nevertheless, it would appear that in the construction adopted by Rotors, the increased stator loss as above was considerably less than the 'parasitic' loss resulting from the variation of airgap permeance. Also, the use of increased number of slots resulted in an improvement of the

---

[12]Inserting the pre-wound coils of the stator winding from the back of the stator and making up the yoke portion by fitting an external magnetic cylinder of 'appropriate' (radial) thickness of previously assembled stack of laminations.

airgap mmf waveform. Rotors' practice, however, did not find much favour with future industrialists.

Further Theoretical Investigations

Further theoretical investigations continued to be carried out and reported by many researchers from around the world, notably by Copeland and Slemon[16,17], Larionov[32], Miyairi and Katakoa[11,13,33], Robertson and Zaky[15] and Rahman[34][13] to explain various aspects related to analysis of hysteresis motor operation and performance dealing with topics such as "analysis of hysteresis motor considering eddy current effect", "a basic equivalent circuit of the hysteresis motor", "airgap flux density distribution in hysteresis motors", "an analysis of the hysteresis motor: parasitic losses[14]", based on several assumptions like constant rotor relative permeability, elliptic shape of the hysteresis loops, 'uniform' airgap (that is with closed stator slots and negligible effect of varying permeance) and 'sinusoidal' distribution of stator conductors.

An analysis of the *radial-type* hysteresis motor was presented by Jaeschke[35] based on the assumptions of a sinusoidal mmf distributed in the airgap and 'infinite' permeability for the stator material and the rotor core. The expression for the "slip force" – force responsible for torque – and driving torque (shown to be independent of speed) were derived in terms of the rotor flux density, obtained by applying a sinusoidal magnetising force to the B-H loop and considering only fundamental component of the resulting flux density waveform, and the applied mmf. The mmf space harmonics were ignored, except for their effect in modifying the speed-torque characteristic resulting from the eddy-current torques. Jaeschke applied the theory to experimental models and inferred that considering an elliptic loop in comparison to an actual loop the error in the rheoretically calculated value of torque using simplified approach would not exceed 7%.

Langen[36] and Nakamura[37] presented a theory of an ideal hysteresis motor and design of a hysteresis motor, respectively, considering sinusoidal variation of flux density and fundamental component of applied mmf, the B-H characteristic being represented as a parabola or a rectangular hyperbola.

Commercially, the motors continued to be manufactured globally by a number of manufacturers, based on simplified design parameters, incorporating increasing advanced permanent-magnet materials for the rotor annulus, to meet the varying applications in the industry and continue to be manufactured in good number. A few typical designs are discussed later.

---

[13]See also, for example,

M.A.Rahman: Field analysis of polyphase hysteresis motor. IEEE Trans., Vol. PAS-99, No.3, May 1980, pp 1164-1171.

[14]The term "parasitic" loss variously used, esp. by Rahman[34], defines the loss associated with the minor loops, caused by various harmonics that arise mainly from the eddy-current effects.

CONSTRUCTIONAL FEATURES[15]

The essential parts of a hysteresis motor are a stator carrying a distributed winding in slots, excited to produce a rotating magnetic field, similar to a poly-phase induction motor or a single-phase motor with phase splitting, and a *rotor that carries no winding* and essentially comprises a permanent-magnet material annulus, fitted on an arbor and a shaft machined from a non-magnetic material such as stainless steel.

Types of Rotor

Based on the construction of the rotor, hysteresis motors are classified *mainly* as

    (a) circumferential

or

    (b) radial type.

The former is shown in Fig.B1.3(a) in which the rotor comprises a *non-magnetic* arbor, consisting of aluminium or non-metallic material such as tufnol, fitted with an annulus of hard or permanent magnet material, having designed radial thickness and axial length. This results in the flux through the annulus to be essentially circumferential, with little to very little flux entering the arbor[16].

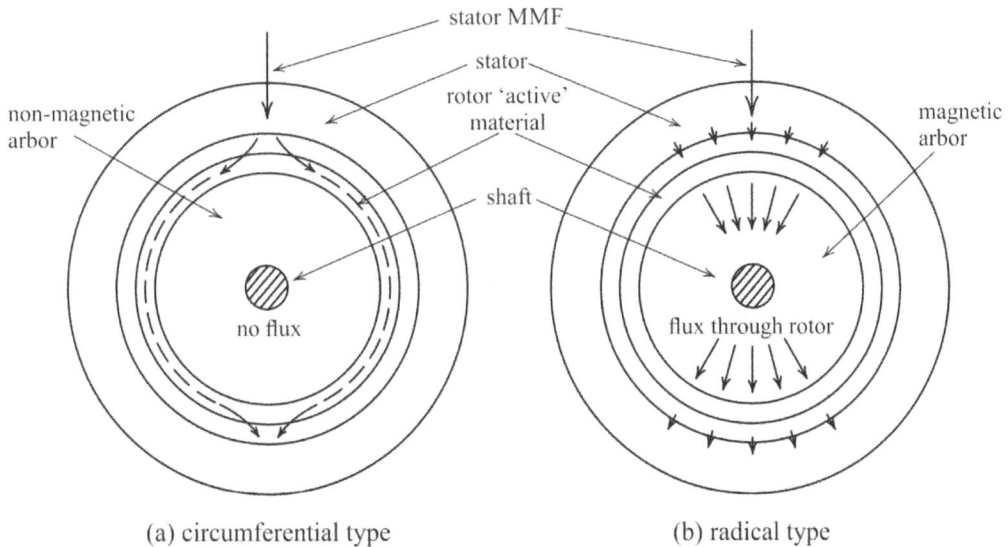

(a) circumferential type   (b) radical type

Fig.B1.3 : Types of rotor for hysteresis motors

---

[15]See, for example,

  S C Bhargava: The hysteresis motor, J.IE(I), Vol.57, Pt. EL 2, 1976, pp 94-98.

[16]However, at increasing excitation as saturation sets in, the permeability of annulus material becomes comparable with that of the arbor material (that is $\mu_r = 1$) and flux 'leaking' into the arbor region. This is discussed in detail later.

The constructional features of a typical circumferential-type motor are depicted in Fig.B1.4.

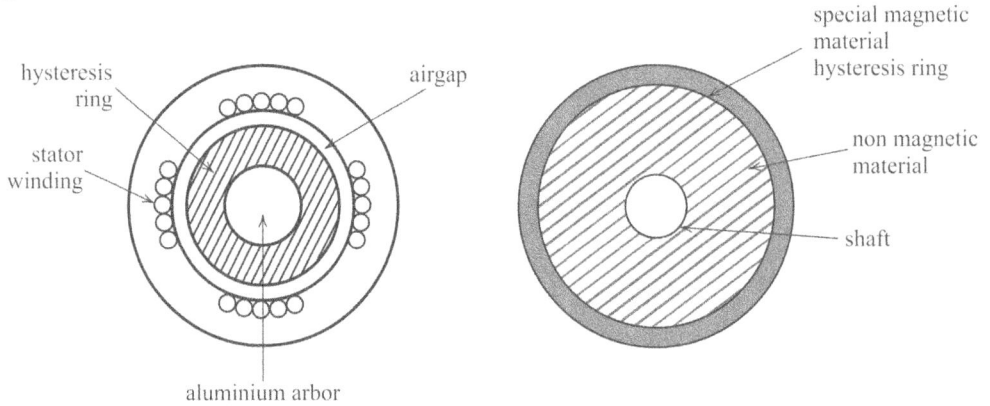

Fig.B1.4 : Cross-sectional details of circumferential-type hysteresis motor

In the second type, the annulus is similar to that in the first type, but the arbor consists of a magnetic material such as 'soft' iron or mild steel having considerably higher relative permeability than that of the annulus. The essential feature in this type is that the flux is mostly directed radially inward as shown in Fig. B1.3(b).

Principle of Operation

An exact theory or principle of operation of a typical hysteresis motor is not possible owing to non-linear and multivalued B-H relationship of the rotor material even when unsaturated. Nevertheless, various researchers have come out with some plausible analyses based on many assumptions as discussed previously. It is, however, known that the operation of the motor and development of torque in it is directly related qualitatively as well as quantitatively on the hysteresis loss (per unit volume or mass) of the rotor active material.

In simplest terms, the operation of a hysteresis motor can be explained by reference to Fig.B1.5.

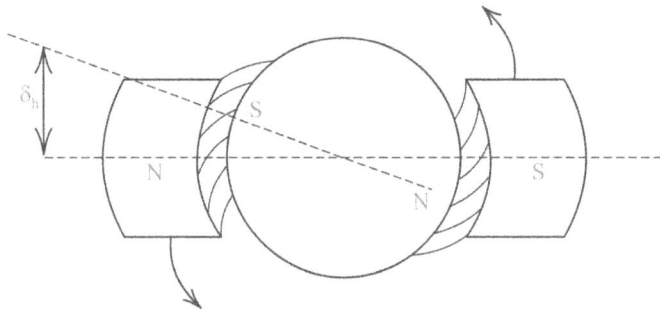

Fig.B1.5 : Simplest explanation of operation of a hysteresis motor

The rotating magnetic field in the airgap due to stator mmf can be visualised to induce poles in the rotor and magnetic field axis as shown that lags the rotating magnetic field axis of the stator by an angle δ, identified as the "angle of hysteretic advance" resulting from the hysteretic property of the rotor annulus material. For a given load condition, the angle δ 'adjusts' itself so that the rotor continues to run at synchronous speed, same as the synchronous speed of the rotating magnetic field of the stator.

A simple, qualitative distribution of flux within the motor is illustrated in Fig.B1.6.

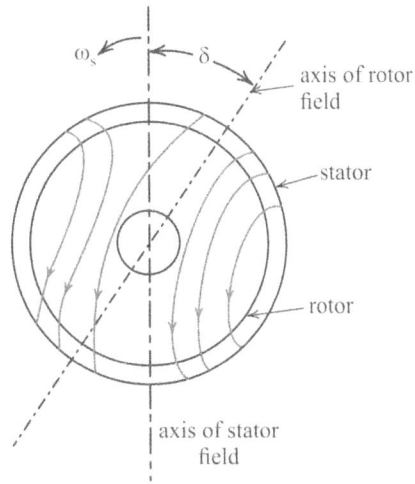

Fig.B1.6 : An approximation of flux distribution in a hysteresis motor

Qualitative Theory

The production of operating torque in a hysteresis motor can be explained qualitatively in terms of the applied rotating field produced in the stator and the resulting **rotor magnetisation**[17] or the magnetic field due to hard magnetic material which has a spatial phase displacement with respect to the former owing to hysteresis property of the annulus material.

A pictorial view of the two fields is depicted in Fig.B1.7 showing the space phase shift relative to each other with the rotor magnetisation lagging behind by the angle, δ.

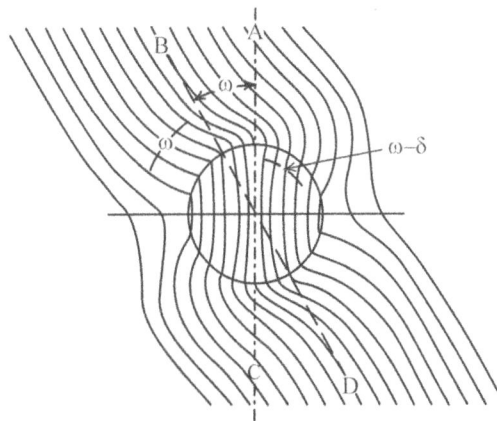

Fig.B1.7 : Magnetising field due to stator and rotor magnetisation

---

[17]The concept of "rotor magnetisation" in the annulus is dealt with in detail later. See Chapter C4.

Assuming the magnetising field due to stator to be varying sinusoidally, or the fundamental component of the actual mmf waveform being usually of 'stepped' form, as plotted in Fig.B1.8, curve A, the resulting waveform of induced flux density in the rotor, curve B, would be distorted and non-sinusoidal, owing hysteresis as derived graphically.

Due to the uniform rotation of the mmf wave, the phenomenon depicted in Fig. B1.8 can be regarded as representing time phase relation between H and B from any reference datum or spatial phase shift at a given instant, the latter being of interest in explaining operation of the machine. The actual wave of B variation can be resolved in a fundamental, curve C, and space harmonics, the former lagging the mmf wave by the angle δ. Then in simple terms the developed torque in the rotor would be proportional to the product of the fundamental component of the stator mmf and the fundamental component of the magnetic field times sin δ[18][31,16].

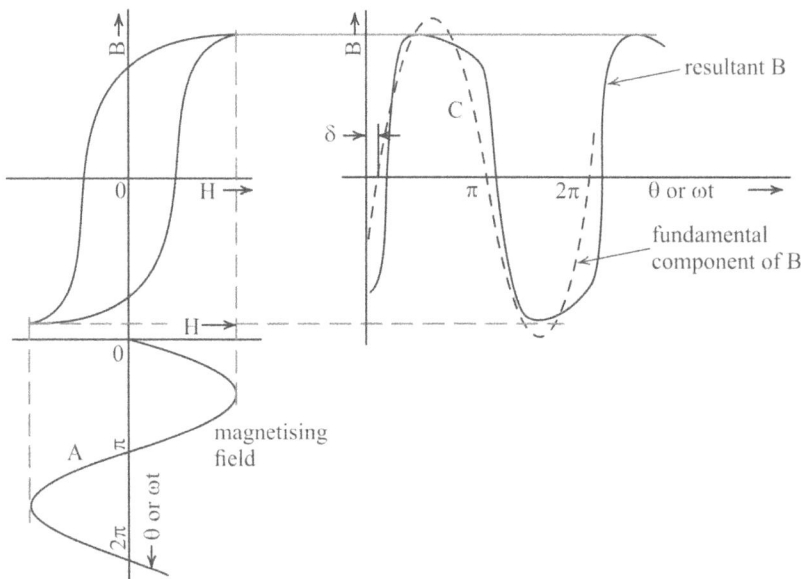

Fig.B1.8 : Sinusoidally varying stator magnetising field and resulting flux density in the rotor

PERFORMANCE AND APPLICATIONS

Torque-speed Curve

The torque-speed curve of a typical hysteresis motor is given in Fig.B1.9. Since the developed torque at any stage of operation numerically equals the hysteresis loss per

---

[18]In practice, the stator mmf would contain numerous harmonics due to the stator slots, usually open to facilitate construction of winding, resulting in production of spurious rotor loss and reduction of net torque[34].

cycle, the ideal torque-speed is shown by the solid line in the figure. Clearly, for different excitations and load conditions, the developed or available torque would be different, but the value of torque remains constant from starting to synchronous speed.

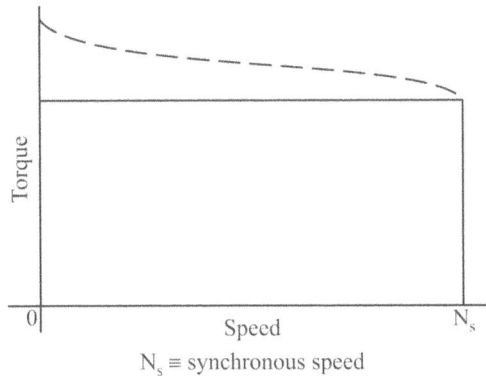

Fig.B1.9 : Torque-speed curve of a hysteresis motor

In practice, however, the shape of the torque-speed characteristic is modified as shown by the dotted curve in the figure. On account of the eddy currents induced in the arbor (in a radial machine) and other metallic parts, there would be excess torque produced during sub-synchronous speeds[19], and the starting torque may be as much as 140% of the 'normal' running or synchronous-speed torque[20,21].

Performance

The power factor of commercial fractional horse power hysteresis motors may usually be low (less than 0.7) and the efficiency less than about 60%, although efficiencies as high as 80% were claimed for a ¼ hp by Rotors[31], the poor efficiency being typically due to reduction of the developed torque due to various factors. Typical terminal characteristics of a small motor are given in Fig.B1.10 showing the variation of input current, developed torque, efficiency and power factor with the stator voltage.

---

[19]Proportional to slip being equal to 1 at starting and 0 at synchronous speed.

[20]It is to be noted that the torque at synchronous speed is capable of being reversed and the motor run as a hysteresis generation or a brake.

[21]See, for example,

D. O'Kelly: (In correspondence), Proc. IEE, Vol.124, No.9, 1977, p 792.

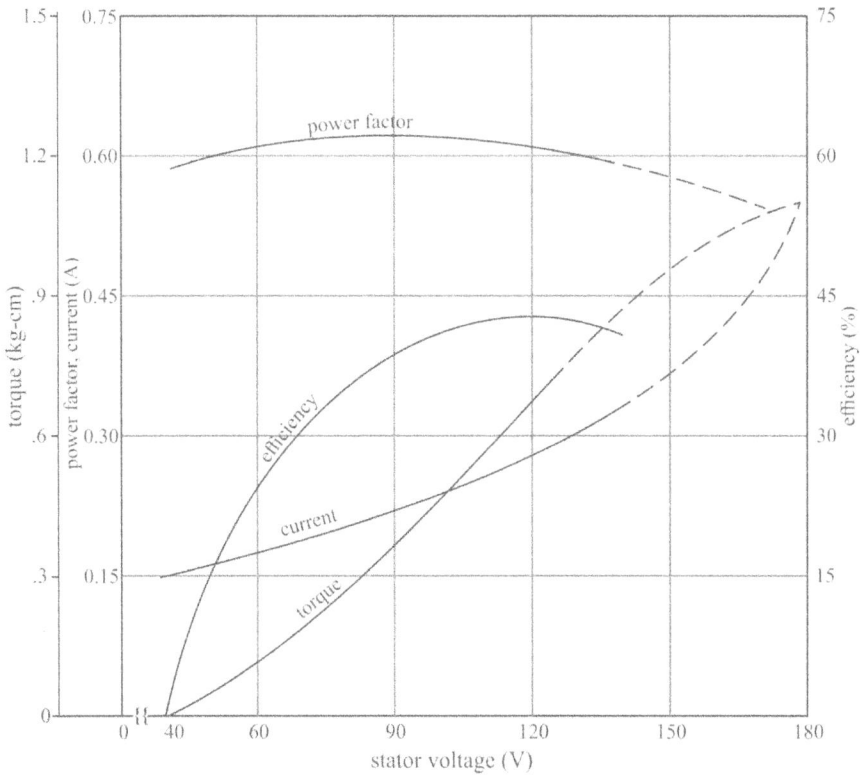

Fig.B1.10 : Performance curves of a small 2-phase hysteresis motor

Advantages and Disadvantages

Advantages

The main advantages of a hysteresis motor are

- The only self-starting synchronous motor where applicable
- As no teeth and no winding are present in the rotor, no mechanical vibrations take place during its operation
- Its operation is quiet and noiseless as there is no vibration
- It is suitable to accelerate inertia loads
- Multispeed operation can be achieved by employing gear train or varying stator supply frequency

Disadvantages

- A hysteresis motor has poor output: nearly one–quarter of output of an induction motor having almost same dimensions
- The motor has relatively low efficiency

- In general, the developed torque and power factor is low
- Hysteresis motors are available in small size only

Applications of Hysteresis Motors

The unique feature of high starting torque up to synchronous speed makes this motor particularly suitable for applications involving high-inertia loads to be started 'instantly' and run synchronously. The earlier use of these motors was confined to driving record players, electric clocks and timers[28] which require sub-fractional horse power machines and *noise-free* operation, and continue to be used still for small clocks. The later, more important applications, included high-quality tape recorders (of spool type) of multi-speed and multi-track design; other analogue recording equipment; gyros; signal apparatus for traffic control; strip chart recorders; automatic winding machines; programme switches for defrosters; washing machines; dish washers; vending machines and computer tape drives comprising 'heavy' components such as magnetic discs or drums. Other forms of hysteresis machines would find their use as "hysteresis clutch" (for example, in nuclear reactor drive uses), hysteresis coupling and brake[22]. Often, speed control in such applications would be possible using some of the conventional methods on the stator side similar to that in a polyphase induction motor.

Due to special nature of construction, hunting is practically non-existent at synchronous speed thus making them ideally suited for application in record players and other recording equipment.

Some Conventional Hysteresis Machines

A. Multi-speed hysteresis motors

For commercial reasons and applications, these motors are designed in three types having two-, four- and six- poles configurations, operating at 50 Hz supply frequency with the synchronous speed being 3000 rpm, 1500 rpm and 1000 rpm, respectively.[23]

The 'basic' motor is wound for two or three phase, but the motor could be operated on a single-phase supply by phase shift obtained by means of a capacitor as shown schematically in Fig.B1.11.

---

[22]To be discussed later in the book.

[23]The motor shaft could be coupled to an appropriate gear box to obtain a wide range of *low* speeds and consequently higher torque.

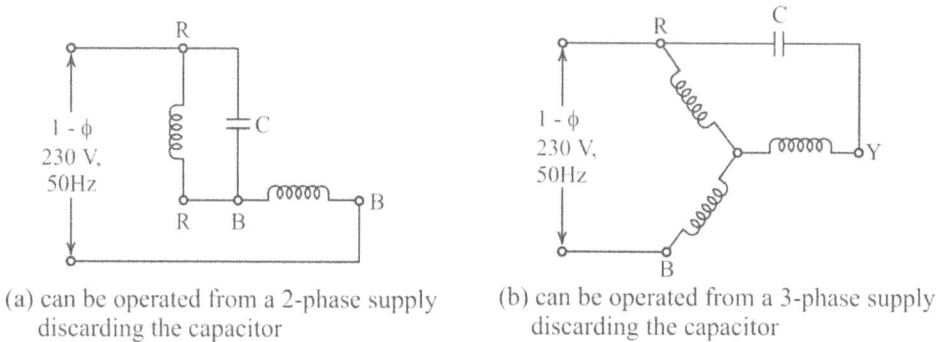

(a) can be operated from a 2-phase supply   (b) can be operated from a 3-phase supply
discarding the capacitor   discarding the capacitor

Fig.B1.11 : Schematic of wiring diagram for the hysteresis motor
for single-phase operation

Construction

The stator core is built from low-loss silicone steel punchings[24], wound using enamelled copper wire using "back-winding" process, and suitably impregnated. The rotor comprises an annulus of hard magnetic material mounted on an aluminium arbor[25]. The rotor assembly is inserted into the stator on sintered bronze bush bearings, affording a small airgap.

Characteristics

Some of the key operating characteristics of the three models of the motor are tabulated in Table B1.1.

Table B1.1: Operating characteristics of hysteresis motors

Type	Input power, W	current, mA	Synchronous speed, rpm pull-out	Starting and torque, g-cm	Efficiency, %[26]
2 P-2 φ	6.5	30	3,000	35	15
4 P-3 φ	6.0	30	1,500	40	10
4 P-2 φ	8.0	40	1,000	26	3.2

B. A hysteresis motor for "rate gyro"

The size of this motor, a 2-pole design, is about 25 mm in diameter and 1 mm in length. The supply for the motor consists of 24 volt, AC at 40 Hz frequency, the synchronous speed being 24,000 rpm[27]. The rotor annulus comprises 35% cobalt steel.

---

[24]Of 'closed'-slot type, 4 slots/pole/phase.

[25]Thus making the motor of circumferential-flux type.

[26]Recall the paper by M A Rahman for factors responsible for low efficiency.

[27]The speed esp. suited for use in a gyro.

C. A hysteresis motor for clockwork

This is a small motor designed for use in a clock, see Fig.B1.12. The main features of the motor are:

rated voltage: 24, 120, or 240 V

rated frequency: 50/60 Hz

power consumption: 5 W

speed: 1-6 rpm (using an in-built gear train)

Fig.B1.12 : A small hysteresis motor for clockwork

A Novel Design

This design envisages primarily a "radial"-type motor using disc-shaped rotor annuli of a thickness of 1 to 2 mm each[28], arranged 'axially' one above the other and separated by 'air'. The stator would be a wound-type in various slots or even shaded-pole salient-pole construction foe single phase operation. The multiplicity of discs on the rotor would provide a simple means to achieve desired power output or developed torque. Such a motor could ideally suit a tape recorder or a record player[29].

Factors Affecting Operation and Performance of Hysteresis Motors

- No of poles: the output or developed torque may not be proportional to the pole pairs
- Airgap: this is the most critical design parameters; can be neither too small nor large; in many cases mechanical aspects, including concentricity, may be more important
- The choice of rotor active material, its B-H characteristics, specific resistance and satisfactory heat treatment as recommended by the material manufacturer
- The radial or circumferential type and whether the arbor is metallic or non-metallic in the case of the former

---

[28]Once again, using vicalloy as the hard magnetic material with the constraints of availability, machining or forming and followed by satisfactory heat treatment.

[29]These players are back in demand to play the poly-vinyl records that were popular a few decades ago.

2 : The Hysteresis Coupling

2

The Hysteresis Coupling

What is a hysteresis coupling?

A hysteresis coupling is characterised by

- a driving 'member' consisting of an electromagnet with salient poles carrying windings excited by DC to produce alternate north and south poles, called the field system

and

- a driven 'member' consisting of a hard or permanent-magnet material in annular or disc form, usually the latter, identified as a rotor as it rotates in the field of the electromagnet.

Two examples of a coupling having an annular rotor are sketched in Fig.2.1(a) whilst schematic of a disc-type coupling is given in Fig.B2.1(b).

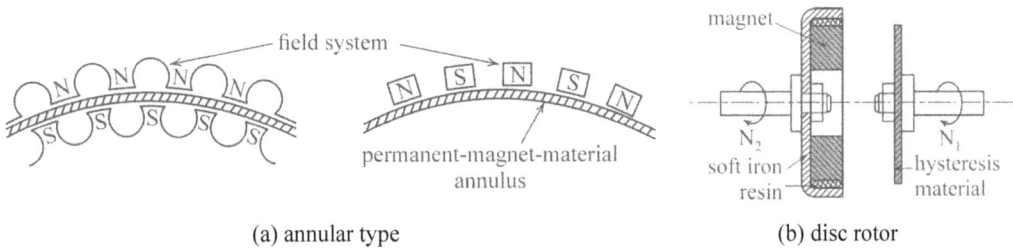

(a) annular type   (b) disc rotor

Fig.B2.1 : Examples of hysteresis coupling

A photographic view of an actual small coupling is given in Fig.B2.2.

Properties/Advantages of Hysteresis Couplings

- free of wear due to non-contact of parts
- low-maintenance
- constant torque at all speeds
- noiseless operation

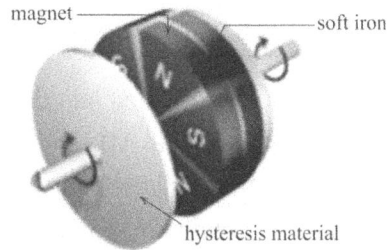

Fig.B2.2 : An actual small hysteresis coupling

These characteristics make such couplings particularly suitable for many position and velocity controls such as

- nuclear reactors
- tape drives
- in industry, for
    - capping machine (bottle closures, screws)
    - labelling machines
    - taper tension and speed control

Driving Torque

The driving torque in a hysteresis coupling is developed in the driven member by virtue of hysteresis loss in the entire volume of the disc or annulus and depends on the excitation to the field system, but is independent of speed and angular position. A typical torque-excitation characteristic of a hysteresis coupling is shown in Fig.B2.3[1].

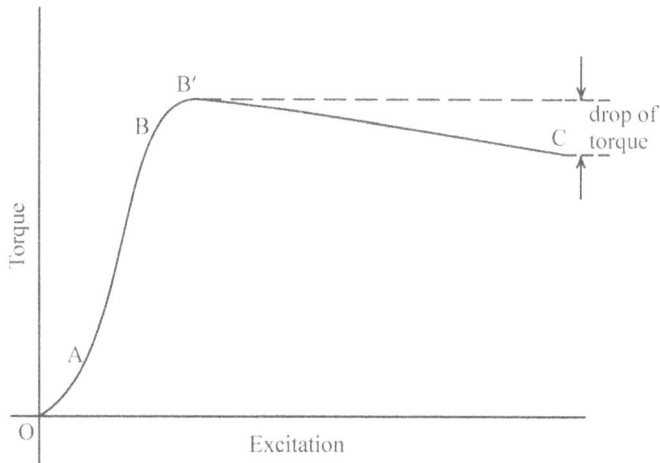

Fig.B2.3 : Torque-excitation characteristic of hysteresis coupling

The curve may be divided into three parts: a curved lower section OA, a near 'linear' normal operating section AB and the upper section BC corresponding to high to very excitations reflecting intense saturation of the hard magnetic material. The developed torque reaches a maximum, point B′, showing no gain in terms of increased excitation;

[1]In practice, rotational hysteresis may contribute to some developed torque, esp. when the material is not much saturated, and torque due to induced eddy current at non-synchronous operation unless the material possesses very high resistivity to induced currents.

in fact, the torque may *drop* at (much) higher excitation in practice, indicated by the dotted line in Fig.B2.3. This may be on several counts like, for example, absence of rotational hysteresis. Note that because of the "power balance", the torque-excitation curve of Fig.B2.3 would resemble the hysteresis loss *per cycle* of the rotor material vs. excitation as derived from the hysteresis loop(s).

PRINCIPLE OF OPERATION

Design aspects of hysteresis couplings have been published in the past[2]. However, little information is available about the mechanism of development of torque in the coupling and its operation as related to magnetic field distribution in the *airgap* and arbor, or inner, regions of the machine and rotor magnetisation. This is mainly on account of highly non-linear and multi-valued properties of the rotor material, both the B-H characteristic and the B-H loop(s), even when not fully saturated, and a lack of near impossible rigorous mathematical analysis of the machine.

Experimental Hysteresis Coupling

Recourse was made to design and develop an experimental hysteresis coupling[3] with the object to obtain a comprehensive magnetic field distribution in the regions of interest of the machine and explain its operation in terms of the components of magnetic field and their interaction. The machine comprised two main parts.

- A field system having two mild steel salient poles fitted to a circular mild steel yoke, carrying windings designed to provide extreme saturation, similar to any salient 2-pole DC machine. The 'axial' length of the field system was 25.4 mm (1 inch) and the active (inner) surface lied on a circle of 82.5 mm diameter.

- A rotor consisting of an annulus fabricated from vicalloy sheet of 0.4 mm thickness, having high resistivity, resulting in negligible induced currents,

---

[2]See, for example,

Yu. A. Yermolin: The magnetic field in the airgap of a hysteresis coupling with a cylindrical rotor, Elektrichestvo. No.8. 1971, pp 63-66.

[3]Strictly a "hysteresis machine" dealt with comprehensively, in full detail, in Part C: Experimental Hysteresis Machine and Analysis. The various sections described here are extracted from Part C, in the context of a detailed discussion pertaining to hysteresis coupling(s), to maintain 'continuity'. See also

S.C.Bhargava: Non-synchronous Operation of a Hysteresis Machine, Ph D Thesis, The University of Aston in Birmingham, UK, 1972.

mounted on a finely machined Perspex arbor, in turn fitted on to a polished stainless steel shaft. The rotor, too, had an axial length of 25 mm, matching that of the field system and an external diameter of 82.2 mm, leaving an airgap of 0.15 mm.

An exploded sketch showing the two parts of the machine is shown in Fig.B2.4 whilst a *to-scale* cross section of the field system and rotor is given in Fig.B2.5 which also shows the details of set of search coils provided to measure flux density distribution in the machine, discussed later.

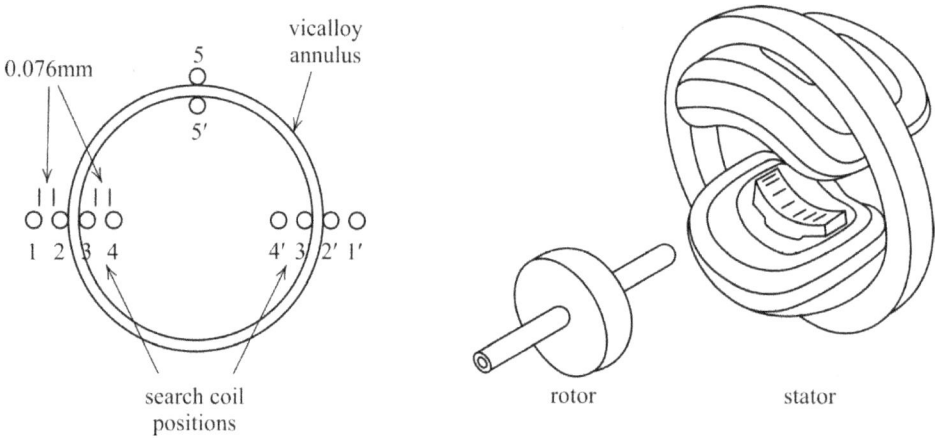

Fig.B2.4 : Sketch of an exploded 2-pole hysteresis coupling

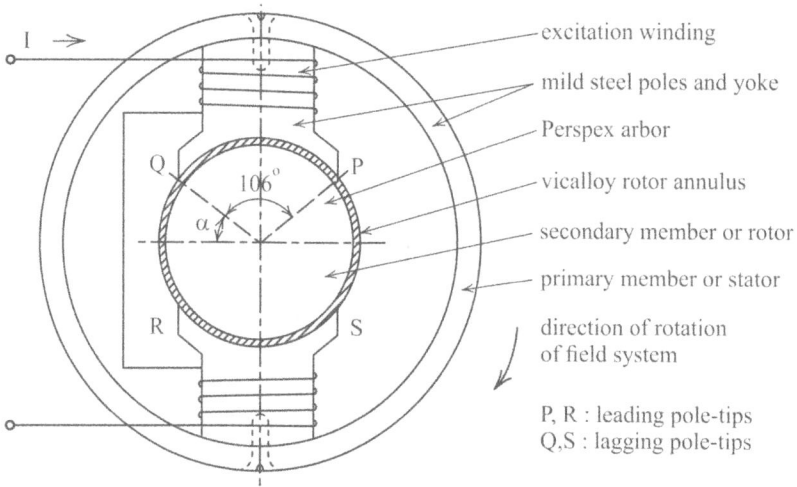

excitation winding

mild steel poles and yoke

Perspex arbor

vicalloy rotor annulus

secondary member or rotor

primary member or stator

direction of rotation
of field system

P, R : leading pole-tips
Q,S : lagging pole-tips

(a) stator and rotor assembly

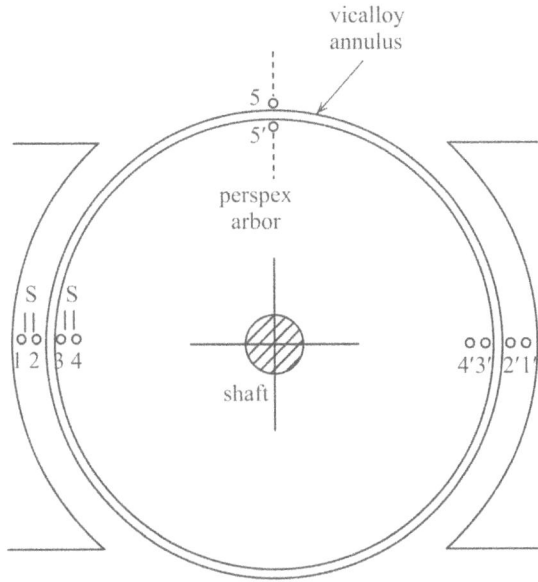

1-1', 2-2' : coil in the air gap
3-3', 4-4' : coil on the arbor side
5-5' : coil surrounding the annulus
S : separation between conductors (49 SWG, enamelled
wire) = 0.076 mm

(b) details of search coils on the rotor

Fig.B2.5 : Cross section of experimental hysteresis coupling and location of search coils on the rotor

Experimental Results

The full-pitch coils on the rotor shown in Fig.B2.5 were used to obtain radial flux density, B_r, variation around the rotor on its external and internal surfaces at various excitations, a typical variation at 0.3 A excitation being shown in Fig.B2.6.

At all excitations, the flux density waveforms are characterised by 'huge' peaks at *lagging* pole tips[4], P and R, as indicated in the figure as a direct consequence of spatial hysteretic property of the rotor active material and resulting from the flux concentration at these pole tips, the difference or unbalance of the peaks at the two pole tips being in direct relation to the developed torque at any excitation.

[4]Clearly this would depend on the direction of rotation and assumption of a reference pole tip.

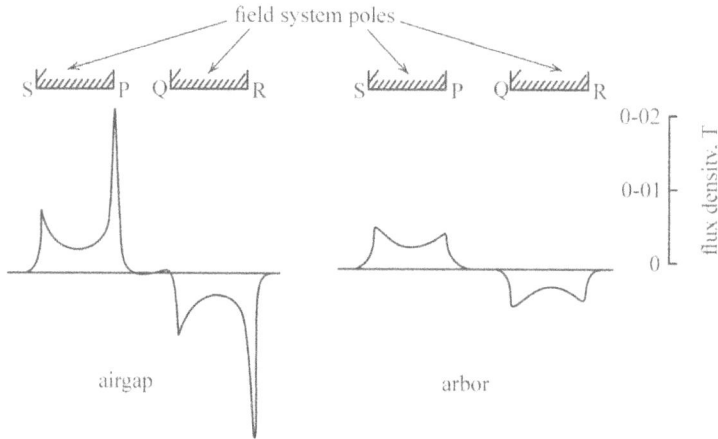

Fig.B2.6 : Radial flux density waveforms at 0.3 A excitation

Analytical Basis

Magnetic scalar potential

In the airgap and arbor regions[5], the magnetic scalar potential, ϕ, is governed by the Laplace equation

$$\nabla^2 \phi = 0$$

and the magnetic flux density in either region is deduced from $-\mu_o \overline{\nabla}\phi$.

Assuming a 2-dimensional variable-separable solution for the Laplace equation in cylindrical coordinates, the radial and peripheral components of the flux density can be written, respectively, as

$$B_r = \sum_{n=1}^{\infty} \left\{ \left(A_n r^{n-1} - B_n r^{-n-1} \right) \cos n\theta + \left(D_n r^{n-1} - E_n r^{-n-1} \right) \sin n\theta \right\}_{n \text{ odd}} \tag{B2.1}$$

and

$$B_\theta = \sum_{n=1}^{\infty} \left\{ -\left(A_n r^{n-1} + B_n r^{-n-1} \right) \sin n\theta + \left(D_n r^{n-1} + E_n r^{-n-1} \right) \cos n\theta \right\}_{n \text{ odd}} \tag{B2.2}$$

where n is the order of harmonic, A_n . . E_n constants, r radial distance from the axis, and θ angular position from an arbitrary datum. Either of the regions has a different set of constants. Equations for the airgap and arbor radial flux density waveforms, B_r, were deduced using eqn.(B2.1) by evaluating the constants (A_n . . E_n) from the Fourier series representation of the search coils waveforms shown in Fig.B2.6. These constants were then used in eqn.(B2.2) to derive the peripheral component of flux density, B_θ; consistent results being obtained by truncating the series for n = 199. Typical B_θ waveforms at excitation of 0.3 A are shown in Fig.B2.7.

---

[5]These *non-magnetic* regions reflect the magnetic field *inside* the magnetic material of the rotor annulus.

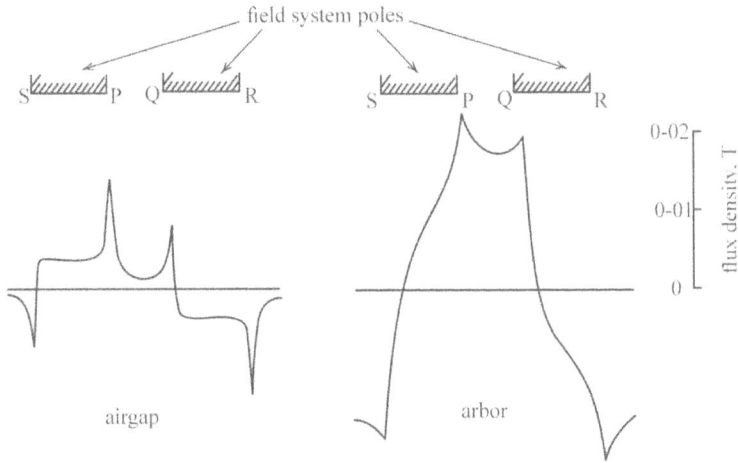

Fig.B2.7 : Peripheral flux density waveforms at 0.3 A excitation

The production of torque in the coupling can be explained in terms of variation of radial and peripheral components of flux density in the airgap and arbor as influenced by hysteresis property of the annulus material[6].

The cardinal feature of the machine, similar to many salient-pole ('DC') machines, is the persistence of the flux concentration *at the pole tips* and into the low-permeability rotor material. This leads to creation of localised surface poles which control the flux distribution in the rotor, to be identified as "rotor magnetisation", and hence the characteristics of the coupling. The rotor magnetisation, a consequence of hysteresis, and considered in detail later, unbalances the pole strengths and there is a net torque.

ENERGY FLOW BY POYNTING THEOREM

The production of torque in a hysteresis coupling can be described by reference to the flow of energy in its airgap and arbor regions and applying Poynting theorem to deduce the power crossing *into* the rotor annulus[7,8].

---

[6]See, for example,

M.J.Jevons and S.C.Bhargava: The salient pole hysteresis coupling, IEEE Trans. on Magnetics, Vol. Mag-11, No, 5, Sept.1975, pp 1461-63.

[7]See, for example,

S.C.Bhargava: Energy flow in a 2-pole hysteresis coupling by Poynting theorem, IEE Proceedings, Vol. 130, Pt. A, No. 6, Sept. 1983, pp 301-305.

[8]The method of deriving torque expression by Poynting theorem has been successfully used earlier by Alger and Oney[38] and Harris *et al*[39] for 3-phase induction motors and Hawthorne[40] for DC and synchronous machines. In all these cases, the analysis (and also the measurement) of developed torque pertains to the energy crossing the airgap.

The fact that the developed torque is independent of speed makes it possible to apply the theorem in its simplest form by defining a suitable Poynting surface and vectors.

Poynting Theorem and Surfaces Pertaining to Rotor Annulus

According to the theorem the surface integral of the normal component of the vector

$$\overline{S} = \text{total } \overline{E} \times \text{total } \overline{H} \tag{B2.3}$$

taken over any *closed* surface equals the total inflow of electromagnetic power to the contained volume; derived from Maxwell's equations, it must apply to any electromagnetic device[9].

Rotor surfaces

To obtain the power flow in the coupling, a Poynting surface is defined, enclosing the rotor annulus as shown in Fig.B2.8.

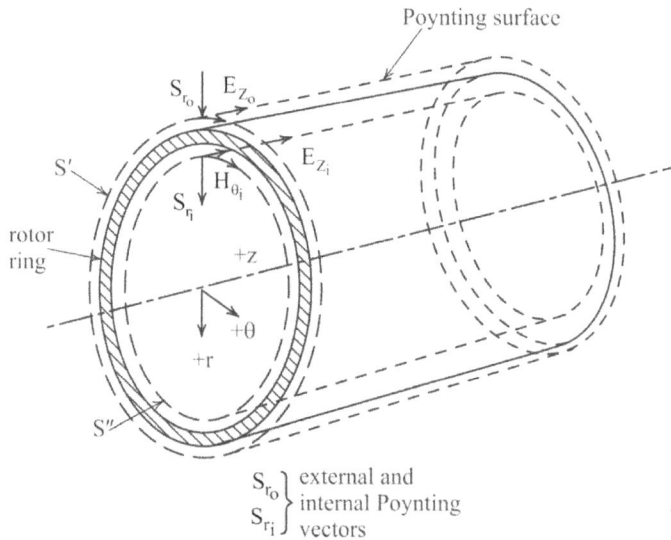

Fig.B2.8 : Poynting surface and vectors pertaining to the rotor annulus

Only the power flow in the radial direction is considered and the two Poynting vectors on the external and internal surfaces are defined by

$$\overline{S}_{r_o} = \overline{E}_{z_o} \times \overline{H}_{\theta_o} \quad \text{and} \quad \overline{S}_{r_i} = \overline{E}_{z_i} \times \overline{H}_{\theta_i} \tag{B2.4}$$

where $\overline{E}_{z_o}$ and $\overline{H}_{\theta_o}$ and $\overline{E}_{z_i}$ and $\overline{H}_{\theta_i}$ denote the resultant axial electric field and peripheral magnetising field vectors on the aitgap (or outer) and inner side of the rotor

---

[9]See, for example,

J.H.Poynting: On the transfer of power in the electromagnetic field, Philos. Trans. R. Soc., Vol. 175. Pt. II, 1884, pp 343-61.

annulus, respectively.   Both S_{r_o} and S_{r_i} are directed *radially inward* as shown, and indicate the flow of power *into* the annulus from the airgap and out to the inner nonmagnetic arbor region. The net power entering the rotor is given by

$$P = \int_{S'} \overline{S}_{r_o} \cdot d\overline{s} + \int_{S''} \overline{S}_{r_i} \cdot d\overline{s} \qquad (B2.5)$$

with S' and S'' as defined in Fig.B2.8[10].

Now E_z is related to B_r, the radial component of flux density, on either surface by the relation $\overline{E} = \overline{v} \times \overline{B}$ and hence the net power crossing *into the annulus* can be expressed as

$$P = K\left\{\int \left(v_{\theta_o} B_{r_o} H_{\theta_o}\right) ds - \int \left(v_{\theta_i} B_{r_i} H_{\theta_i}\right) ds\right\} \qquad (B2.6)$$

where K is a constant. Then, assuming an arbitrary relative speed of rotation of ω rad/s (related to v above), the developed torque in Nm can be written as

$$T = P/\omega \qquad (B2.7)$$

Eqn.(B2.6) points to an important aspect of the condition of production of torque in a hysteresis machine. If the variations of B_r and H_θ were symmetrical about pole and interpolar axes, respectively, as in a simple symmetric magnetic circuit, or indeed in the coupling under 'static' condition, the net power flow would be zero and hence no torque developed. In the hysteresis coupling, the hysteretic property of the rotor annulus material results in a spatial shift of the magnetic field after magnetisation of the rotor annulus leading to nonzero power flow. This is also reflected in the spatial variation of B_r and B_θ waveforms shown in Figs.B2.6 and B2.7.

Another point which has direct bearing on the location of probes in the rotor, as shown in Fig.B2.5, for the measurement of energy flow through the airgap is the non-dependence of the expression for total power on the radius at which measurements were made. This is to be expected since no net power is absorbed in the air region of the machine.

Energy Flow in the Experimental Coupling

The schematic of the experimental coupling, depicted variously in Figs.B2.4 and B2.5, showing the drive mechanism is shown in Fig.B2.9 whilst its photographic view is given in Fig.B2.10.

---

[10]Although only the total power, that is the sum of integrated values of $\overline{S}_{r_o}$ and $\overline{S}_{r_i}$ on S' and S'', respectively, is a meaningful quantity, a useful interpretation can be given to the 'power density' at a *point* (the vectors $\overline{S}_{r_o}$ and $\overline{S}_{r_i}$) as discussed later.

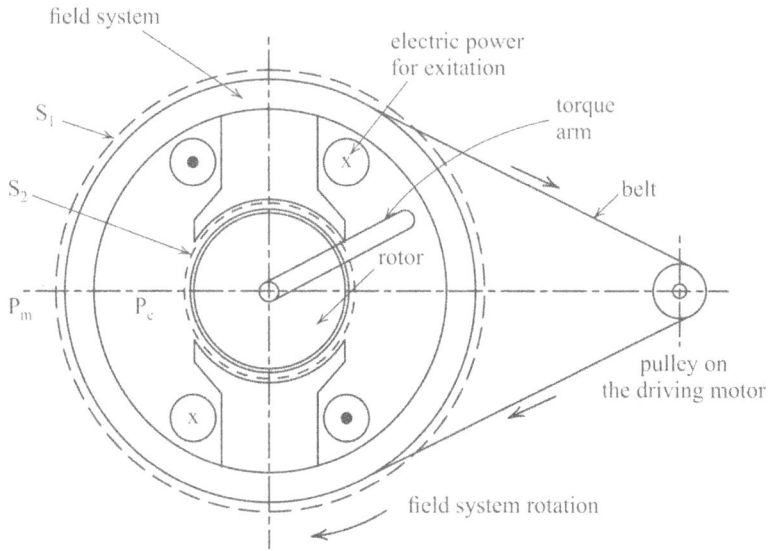

Fig.B2.9 : Schematic of the experimental coupling showing its construction and drive mechanism

Fig.B2.10 : Photographic view of the experimental coupling

The active parts of the coupling are mounted on a rigid platform with the axis of rotation of the field system being *vertical*. The rotor assembly is held stationary, fitted with a "torque arm"[11], and is prevented from rotation at any stage for ease of making various measurements. The field system is rotated from a Velodyne drive through a belt at a speed of 50 rpm.

Observe that the power crossing the airgap is delivered by the driving motor (the Velodyne), with the role of the field winding on the poles being that of providing only the magnetisation of the annulus, resulting in hysteresis loss and production of torque. Qualitatively, when electrical power is supplied to the field system in the form of DC excitation, energy crosses the airgap towards the rotor annulus, with a part of this being 'used up' as hysteresis loss, equivalent to the developed torque, and the balance 'returning' to the driving motor via the rotor arbor region, the rotor shaft, the gantry plate and other mechanical structure of the assembly (see Fig.B2.10)[12].

Transference of power

With reference to Fig.B2.9, let P_m be the net mechanical power flow from the drive motor through S_1 and P_e through S_2. Then, neglecting friction and windage,

$$P_e = P_m \tag{B2.8}$$

Let the hysteresis loss per revolution be W_h Ws (or joule). Then the total loss per second will be NW_h W, where N is the speed of rotation of the field system in rev/s. As P_e is responsible for hysteresis loss in the rotor

$$P_e = N\, W_h = P_m \tag{B2.9}$$

The hysteresis loss P_e is observed as torque T exerted on the rotating field system because of the relative motion. Hence, if ω is the angular speed of rotation in rad/s

$$P_m = \omega\, T$$

so that

$$T = P_m/\omega = (N\, W_h)/\omega \tag{B2.10}$$

or, putting $N = \omega/2\pi$,

$$T = W_h/2\pi \ \text{Nm or} \ W_h \ W_s(\text{syn}) \tag{B2.11}$$

This shows that the developed torque is numerically equal to the hysteresis loss in the rotor active material per revolution and is independent of speed – a fact that was first noted by Steinmetz[13].

---

[11]A flat stainless steel bar about 30 cm long and a section of 12 mm × 4 mm, provided with strain gauges on opposite flat surfaces at mid-length. The bar is held so as to get strained when torque is developed in the rotor, sensed by the gauges. The latter are connected to form a 'bridge'. The out-of-balance output of the bridge being proportional to the developed torque is fed into a calibrated 'output meter' for measurement.

[12]There would be some local 'exchange' of energy across the airgap at the leading pole tips that may result in loss of developed torque at higher excitations.

[13]See, for example,

C.P.Steinmetz: Theory and Calculation of Electrical Apparatus (book), McGraw-Hill, New York. 1917, pp 168-71.

Experimentally-derived and measured torque variation

From the calculated values of P_e, the equivalent torque was derived using eqn.(B2.10) for various excitations, the values of B_{r_o} and B_{r_i} having been used from the outputs of search coils detailed in Fig.B2.5 and shown in Fig.B2.6 for the two regions at 0.3 A excitation. The derived and measured torque values are compared in Table B2.1.

Table B2.1: Calculated and measured torque against excitation

Excitation, A	Torque, Nm		
	derived	measured	theoretical
0.1	0.008	0.0048	0.001
0.15	0.019	0.015	0.016
0.2	0.05	0.041	0.0355
0.25	0.056	0.045	0.0421
0.3	0.085	0.048	0.0516
0.4	0.058	0.047	linear theory
0.6	0.056	0.045	not
1.0	0.05	0.041	applicable

The measured torque-excitation characteristic is fully realised by deriving the torques using Poynting theorem approach at corresponding excitations from the measured B_r and H_θ variations in the two *nonmagnetic* regions. Even though the derived torque values are relatively much higher than the measured torque at higher excitations (>0.3 A), mainly because the net torque is obtained from the *difference* of total power flows on the external and internal rotor surfaces, these being two almost equal quantities on account of the use of a thin annulus, and unavoidable experimental errors, it is observed that the *trend* of torque variation with excitation is nearly the same as for the measured curve.

Poynting Vector Distribution

A quantity of interest in the present analysis is the product of B_r and H_θ, representing S_r in arbitrary units and its variation round the rotor annulus at a given excitation. The variation can be imparted some interpretation to indicate 'power density' at a particular location along the rotor periphery to give an idea of how the energy flows across the airgap.

As an illustration, plots of S_r for the excitation currents of 0.1 and 0.3 A are given in Fig.B2.11. Because the direction of both E and H reverses under each pole, the plots in Fig.B2.11 apply over either pole pitch and the power flow under the pole arc is always of identical sign.

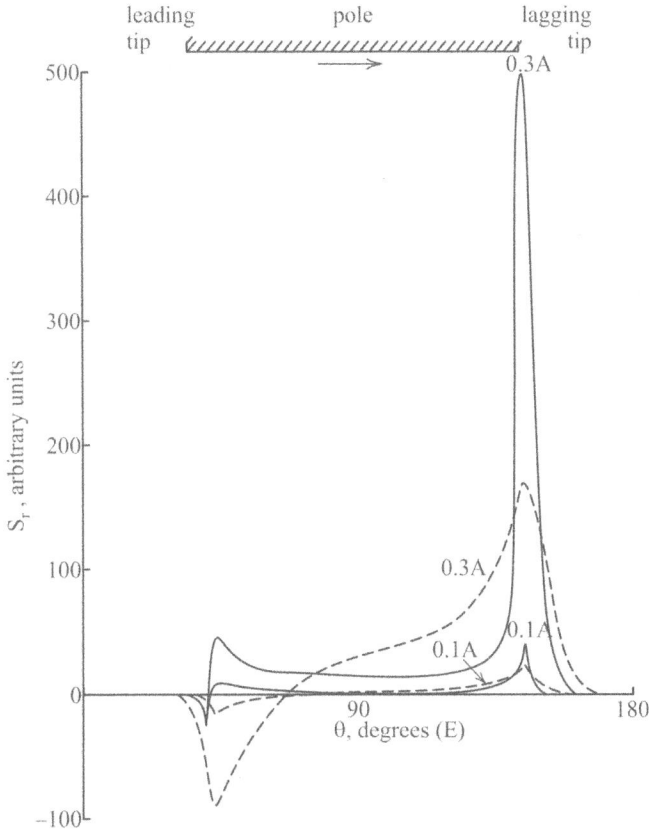

Fig.B2.11: Poynting vector distribution in the airgap and arbor regions of hysteresis coupling at 0.1 and 0.3 A excitation

——————— airgap; --------- arbor

As seen, the magnitude of S_r at either excitation is greatest at the lagging pole tips, suggesting that a large proportion of power flow takes place from these tips. This would be expected as H_θ variation is nearly flat over the pole arc owing to hysteresis and *rotor magnetisation* whilst B_r distribution contains huge peaks near the lagging pole tips (see Fig.B2.6).

The plots have negative values opposite leading pole tips and, if the previous interpretation is assumed to hold, this means that energy *leaves* the rotor annulus to 'return' to the field system.

Torque at Higher Excitations

The loss of torque at higher excitations was explained earlier in terms of the formation of *surface poles* on the external rotor surface adjacent to the pole tips, resulting in consequent reduction in the effective volume of the rotor. In terms of power flow considered here and by reference to Fig.B2.11, the reduction of torque can be attributed to the increasing proportion of energy being returned to the field system at the leading pole tips whilst the magnetic condition of the annulus near the lagging pole tips approaches intense saturation, with no further increase in power flow.

THEORY OF THE COUPLING TORQUE BASED ON TWO-POTENTIAL CONCEPT[14]

A quantitative analysis of predicting developed torque in the coupling, related to its design details and magnetic properties of the rotor annulus material, can be derived in terms of magnetic scalar potential distribution in the three regions of the rotor: airgap, annulus and arbor.

The experimental coupling described earlier is used deriving the torque at various excitations. [See Fig.B2.5 and Table B2.1].

Magnetic Scalar Potential Distribution in the Coupling

Assumptions

- The flux distribution in the machine is not affected by eddy currents at any stage of operation. This is justified since the annulus is made of vicalloy – a material of very high resistivity ($0.7 \ \mu\Omega$-m) – and the arbor being made of Perspex.

- The field distribution is essentially two-dimensional; this holding true since the (axial) length of the machine is very much greater than the airgap.

- The relative permeability of poles and yoke is infinite.

- The relative permeability of rotor material is constant *for a given excitation*. The values of μ_r for the rotor for different excitations are derived from the ballistic magnetisation curve of vicalloy (see Appendix A), reproduced in Fig.B2.12. The derived relative permeability curve is expressed in Table B2.2[15].

---

[14]See, for example,

S.C.Bhargava: Theory of Hysteresis coupling torque. Electric Machines and Electromagnetics, No. 5, 1980, pp 391-405.

[15]*Ibid.*

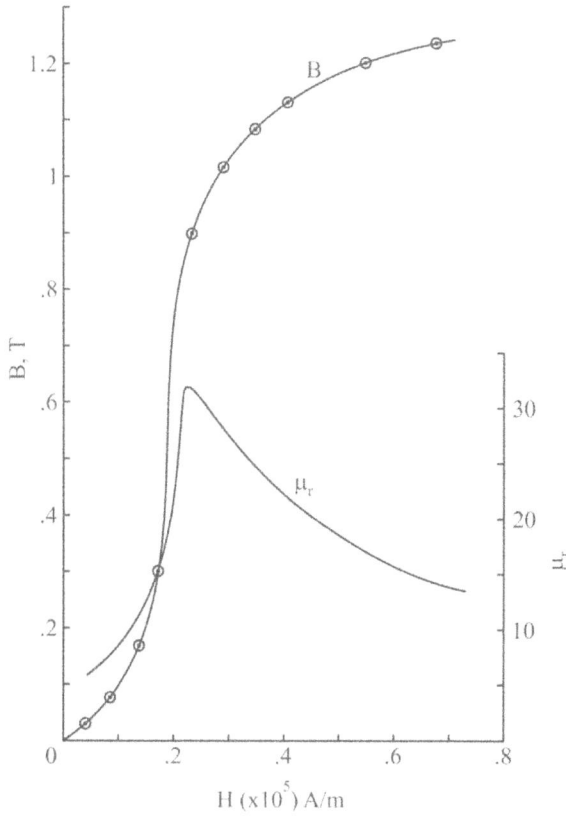

Fig.B2.12 : B-H and (derived) μ_r – H characteristics of vicalloy

Table B2.2: Excitation vs. relative permeability for vicaloy rotor

No of turns /pole: 2200

Excitation current, I A	mmf per pole NI A	magnetising field through the annulus, H $H = 2\ NI/l\ A/m \times 10^5$	rotor relative permeability, μ_r
0.1	220	0.0343	6
0.2	440	0.0686	7
0.3	660	0.1029	9
0.4	880	0.1372	11
0.5	1100	0.1715	17

Derivation of General Torque Equation

Electromagnetic field equations

Under the foregoing assumptions, Maxwell's equations for the magnetic field inside the machine are given by

$$\text{curl } \overline{H} = 0 \tag{B2.12a}$$

$$\text{div } \overline{B} = 0 \tag{B2.12b}$$

Following eqn.(B2.11a), $\overline{H}$ can be represented by a magnetic scalar potential ϕ such that

$$\overline{H} = -\text{grad } \phi \tag{B2.13}$$

Then ϕ would satisfy the Laplace's equation

$$\nabla^2 \phi = 0 \tag{B2.14}$$

which in cylindrical coordinates can be expanded as

$$\frac{\partial^2 \phi}{\partial r^2} + \frac{1}{r}\frac{\partial \phi}{\partial r} + \frac{1}{r^2}\frac{\partial^2 \phi}{\partial \theta^2} = 0 \tag{B2.15}$$

Using the method of separation of variables[16], the scalar potentials ϕ_1, ϕ_2 and ϕ_3 in the airgap, rotor annulus and arbor region, respectively, can be expressed by

$$\phi_1 = \sum_{n=1}^{\infty}\left\{\left(A_n r^n + B_n r^{-n}\right)\sin n\theta + \left(C_n r^n + D_n r^{-n}\right)\cos n\theta\right\}_{n \text{ odd}} \tag{B2.16a}$$

$$\phi_2 = \sum_{n=1}^{\infty}\left\{\left(E_n r^n + F_n r^{-n}\right)\sin n\theta + \left(G_n r^n + K_n r^{-n}\right)\cos n\theta\right\}_{n \text{ odd}} \tag{B2.16b}$$

$$\phi_3 = \sum_{n=1}^{\infty}\left\{L_n r^n \sin n\theta + M_n r^n \cos n\theta\right\}_{n \text{ odd}} \tag{B2.16c}$$

where r is the general radius vector in any of the three regions, θ the angular position of the radius vector from an arbitrary datum such as the interpolar axis of the machine[17], 'n' the order of harmonic in the mmf waveform and A_n, . . . , M_n are constants to be evaluated.

Boundary conditions

Following the usual procedure, the constants A_n through M_n in eqns.(B2.16) can be evaluated by applying appropriate boundary conditions at the two air-vicalloy interfaces of the rotor; these conditions being the continuity of normal component of flux density

---

[16]See, for example,

P.Moon and D.E.Spencer: Field Theory Handbook, Springer-Verlag, Berlin, 1961.

[17]There are no negative power terms in the expression for ϕ_3 since it has to be finite at $r = 0$.

(B) and tangential component of magnetising field (H). Also, on the pole surface, the potential ϕ_1 should equal the potential distribution due to the excitation winding. Thus

■ At $r = r_3$ (pole surface)

$$\phi_1 = \sum_{n=1}^{\infty} \left\{ \left(A_n r_3^n + B_n r_3^{-n} \right) \sin n\theta + \left(C_n r_3^n + D_n r_3^{-n} \right) \cos n\theta \right\}_{n \text{ odd}}$$

$$= \sum_{n=1}^{\infty} \left\{ \phi_n' \sin n\theta + \phi_n'' \cos n\theta \right\}_{n \text{ odd}} \tag{B2.17}$$

In eqn.(B2.17), ϕ_n' and ϕ_n'' represent the coefficients of Fourier expansion of the potential waveform on the pole surface. This variation is shown in Fig.B2.13, derived from to-scale Teledeltos plot[18] and, later, analytically[19,20]. As shown later, the resultant field in the interpolar region follows an inverse Jacobian elliptic relationship.

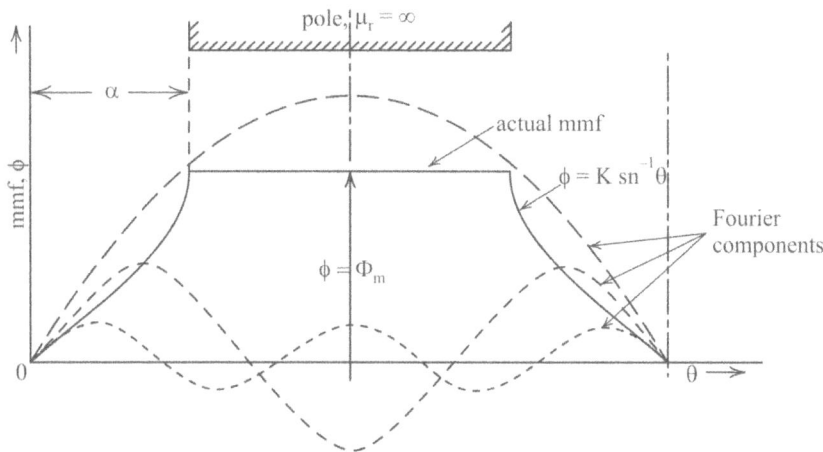

Fig.B2.13: Mmf distribution on the pole surface of the 2-pole hysteresis coupling

Then, for example, with the datum as shown

$$\phi_1 = \sum_{n=1}^{\infty} \left\{ \phi_n' \sin n\theta \right\}_{n \text{ odd}} \tag{B2.17a}$$

■ At $r = r_2$ (external rotor surface), making use of eqn.(B2.13)

---

[18]See, for example,

S.C.Bhargava: Non-synchronous Operation of a Hysteresis Machine, Ph.D. Thesis. The University of Aston, U K, 1972.

[19]*Ibid.*

[20]See, for example,

M.J.Jevons: Magnetic field calculation for a salient-pole hysteresis coupling, COMPUMAG, 1976.

In the airgap

$$H_{1_\theta} = \frac{-1}{r_2}\frac{\partial \phi_1}{\partial \theta} = \sum_{n=1}^{\infty} - n \left\{ \left(A_n \, r_2^{\,n-1} + B_n r_2^{\,-n-1} \right) \cos n\theta + \left(C_n \, r_2^{\,n-1} + D_n \, r_2^{\,-n-1} \right) \sin n\theta \right\}_{n \text{ odd}}$$

(B2.18a)

In the annulus

$$H_{2_\theta} = \frac{-1}{r_2}\frac{\partial \phi_2}{\partial \theta} = \sum_{n=1}^{\infty} - n \left\{ \left(E_n \, r_2^{\,n-1} + F_n r_2^{\,-n-1} \right) \cos n\theta + \left(G_n' \, r_2^{\,n-1} + K_n \, r_2^{\,-n-1} \right) \sin n\theta \right\}_{n \text{ odd}}$$

(B2.18b)

and $H_{1_\theta} = H_{2_\theta}$ (B2.18c)

Also,

$$B_{1_n} = -\mu_0 \frac{\partial \phi_1}{\partial \theta} = -\sum_{n=1}^{\infty} \mu_o \, n \left\{ \left(A_n \, r_2^{\,n-1} - B_n r_2^{\,-n-1} \right) \sin n\theta - \left(C_n \, r_2^{\,n-1} - D_n \, r_2^{\,-n-1} \right) \cos n\theta \right\}_{n \text{ odd}}$$

(B2.19a)

$$B_{2_n} = -\mu_0 \mu_r \frac{\partial \phi_2}{\partial r} = -\sum_{n=1}^{\infty} \mu_o \mu_r n \left\{ \left(E_n \, r_2^{\,n-1} - F_n r_2^{\,-n-1} \right) \sin n\theta - \left(G_n \, r_2^{\,n-1} - K_n \, r_2^{\,-n-1} \right) \cos n\theta \right\}_{n \text{ odd}}$$

(B2.19b)

and

  $B_{1_n} = B_{2_n}$ (B2.19c)

■   At $r = r_1$ (internal annulur surface), making use of eqn.(B.2.13)

$$H_{2_\theta} = -\sum_{n=1}^{\infty} n \left\{ \left(E_n \, r_1^{\,n-1} - F_n \, r_1^{\,-n-1} \right) \cos n\theta - \left(G_n \, r_1^{\,n-1} - K_n \, r_1^{\,-n-1} \right) \sin n\theta \right\}_{n \text{ odd}} \quad \text{(B2.20a)}$$

$$H_{3_\theta} = -\sum_{n=1}^{\infty} n \left\{ L_n \, r_1^{\,n-1} \cos n\theta + M_n \, r_1^{\,n-1} \sin n\theta \right\}_{n \text{ odd}} \quad \text{(B2.20b)}$$

and $H_{2_\theta} = H_{3_\theta}$ (B2.20c)

   Also,

$$B_{2_n} = -\sum_{n=1}^{\infty} \mu_0 \mu_r n \left\{ \left(E_n \, r_1^{\,n-1} + F_n \, r_1^{\,-n-1} \right) \sin n\theta + \left(G_n \, r_1^{\,n-1} + K_n \, r_1^{\,-n-1} \right) \sin n\theta \right\}_{n \text{ odd}} \quad \text{(B2.21a)}$$

$$B_{3_n} = -\sum_{n=1}^{\infty} \mu_o \, n \left\{ L_n \, r_1^{\,n-1} \sin n\theta + M_n \, r_1^{\,n-1} \cos n\theta \right\}_{n \text{ odd}} \quad \text{(B2.21b)}$$

and $B_{2_n} = B_{3_n}$ (B2.21c)

Developed Torque

Symmetrical field distribution

In the foregoing analysis, the hysteresis property of the rotor annulus material has not been accounted and hence, with the flux distribution [given by eqns.(B2.16) through (B2.21)] being symmetrical in the machine, say about the pole axis of the field system, the developed torque would be zero. This can be shown by using a general torque expression, as for example developed by Teare based on the principle of virtual work[21], which for the present two-dimensional analysis (axial length L) is reduced to

$$T = L \int_{r_1}^{r_2} \int_0^{2\pi} \left(B_r \frac{\partial H_r}{\partial \theta} + B_\theta \frac{\partial H_\theta}{\partial \theta} \right) d\theta \, r \, dr \tag{B2.22}$$

Thus, using eqn.(B2.13), for the rotor

$$H_r = H_{2_r} = -\sum_{n=1}^{\infty} n \left\{ \left(E_n \, r^{n-1} + F_n \, r^{-n-1} \right) \sin n\theta + \left(G_n \, r^{n-1} + K_n \, r^{-n-1} \right) \cos n\theta \right\}_{n \text{ odd}} \tag{B2.23}$$

and

$$H_\theta = H_{2_\theta} = -\sum_{n=1}^{\infty} n \left\{ \left(E_n \, r^{n-1} - F_n \, r^{-n-1} \right) \cos n\theta - \left(G_n \, r^{n-1} - K_n \, r^{-n-1} \right) \sin n\theta \right\}_{n \text{ odd}} \tag{B2.24}$$

with the respective flux density components given by

$$B = \mu_0 \mu_r H \tag{B2.25}$$

When the components of $\overline{B}$ and $\overline{H}$ from these equations are substituted in eqn.(B2.22), the developed torque is seen to be zero.

Field distribution accounting for hysteresis

The magnetic field distribution in the machine is modified due to hysteresis property of the rotor annulus material as revealed by hysteresis loops at varying magnetising field or excitation. The hysteresis effect is manifest as *rotor magnetisation* and potential under the action of 'initial' rotation of the field system, the actual spatial hysteresis exhibiting in the form of the resultant radial flux density increasing considerably at the lagging pole tips of the field system with corresponding reduction at the leading pole tips. During extensive experiments it was observed that the *peripheral* flux density *within* the vicalloy annulus reverses near the leading pole tip. This phenomenon is depicted in the waveforms of Fig.B2.14, measured at 0.3 A excitation.

---

[21]See, for example,

B.R.Teare, Jr.: Theory of hysteresis motor torque, Trans. AIEE, Vol.59, 1940, pp 12-15.

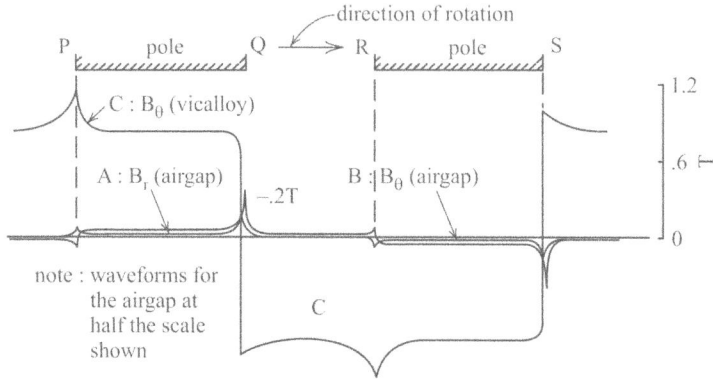

Fig.B2.14 : Radial and peripheral flux density in the annulus and airgap at 0.3 A excitation

These effects are also tantamount to the dynamic flux distribution in the rotor annulus as shown in Fig.B2.15 revealing that under operating condition and by virtue of rotor magnetisation and spatial hysteresis, poles are formed in the annulus which align with the pole tips as shown, leading to production of net torque.

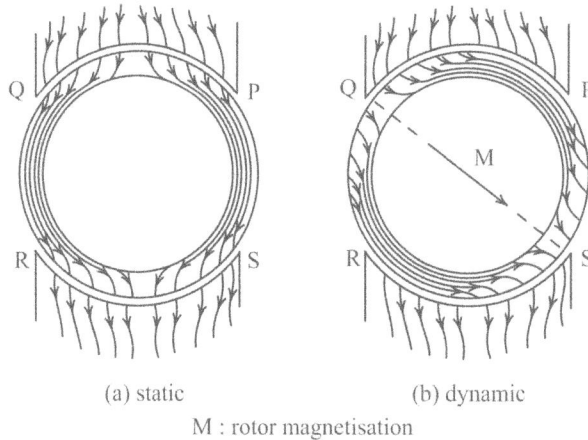

(a) static   (b) dynamic

M : rotor magnetisation

Fig.B2.15 : Rotor magnetisation and 'induced' poles

This leads to three conclusions:

1. The formation of independent poles on the external rotor surface is equivalent to an independent magnetic potential due to annulus hard material which comes into existence ideally following one complete revolution of the field system; in practice, several revolutions may be required to bring the rotor material into cyclic magnetic state.

2. Under the assumption of relatively very little mmf required for the airgap, the amplitude of the 'rotor mmf' would be numerically equal to the maximum value of the stator mmf or ϕ_m as given in Fig.B2.13.

3. Because of the geometry of the field system, esp. with salient poles, the peak of the rotor mmf is always aligned with the lagging pole tip, irrespective of the excitation or the condition of spatial hysteresis due to rotor ring.

It follows that the (scalar) magnetic potential distribution on the external rotor surface would be a 'triangular' wave, similar to the armature reaction in a DC machine as shown in Fig.B2.16 along with the mmf distribution due to main poles. The triangular variation is also evident from constant value of B_θ *in the rotor annulus* under the pole as shown by variation "C" in Fig.B2.14.

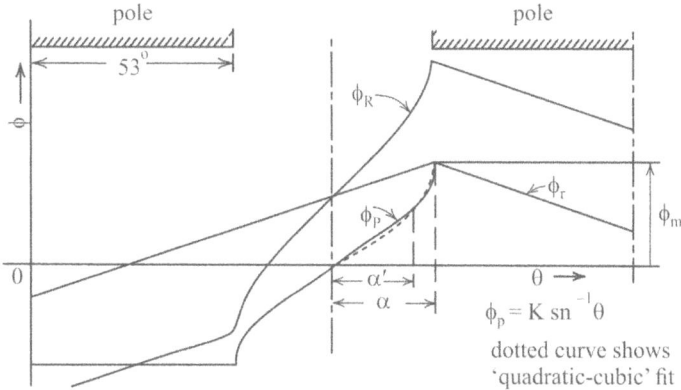

Fig.B2.16 : Mmf distribution due to main excitation and rotor magnetisation

Physically, the presence of a non-zero B_θ due to rotor potential under the main poles would explain the production of a net, useful torque, similar to that in any electrical machine.

Derivation of torque equation considering rotor potential

The magnetic potential in interpolar region due to stator excitation alone, ϕ_p, shown in Fig.B2.16, is given by

$$\phi_p = K\ sn^{-1}\theta \quad A \text{ [inverse Jacobian elliptic relationship]} \qquad (B2.26)$$

where K is a constant. However, to represent the total stator potential by a Fourier series, the interpolar variation is approximated by a 'quadratic-cubic' fit such that

$$\phi_p = a_1\theta + b_1\theta^2 \qquad \ldots \quad 0 < \theta < \alpha' \qquad (B2.27a)$$

$$\phi_p = a_2\theta + b_2\theta^2 + c\ \theta^3 \qquad \ldots \quad \alpha' < \theta < \alpha \qquad (B2.27b)$$

Also, from Fig.B2.16,

$$\phi_p = \phi_m \qquad \ldots \quad \alpha < \theta < \pi/2 \qquad (B2.27c)$$

Since the series would contain only odd sine harmonics, ϕ_p can be written

$$\phi_p = \sum_{1,3}^{n} \phi_n\ \sin n\theta \qquad (B2.28)$$

where ϕ_n is expressed by

$$\phi_n = \frac{\pi}{4}\left[\int_0^{\alpha'}\left(a_1\,\theta + b_1\,\theta^2\right)\sin\,n\theta\,d\theta + \int_{\alpha'}^{\alpha}\left(a_2\,\theta + b_2\,\theta^2 + c\,\theta^3\right)\sin\,n\theta\,d\theta + \int_{\alpha}^{\pi/2}\phi_m\,\sin\,n\theta\,d\theta\right]$$

(B2.29)

Then, from eqns.(B2.17) and (B2.28),

$$A_n r_3^{\,n} + B_n\,r_3^{\,-n} = \phi_n \qquad\text{and}\qquad C_n\,r_3^{\,n} + D_n\,r_3^{\,-n} = 0$$

(B2.30)

With reference to the chosen origin (Fig.B2.16), the rotor magnetic potential is a displaced triangular wave with its Fourier expansion given by

$$\left.\phi_r\right|_{r=r_2} = \sum_{n=1}^{\infty}\left[\phi_{r_n}'\,\sin\,n\theta + \phi_{r_n}''\,\cos\,n\theta\right]_{n\ odd}$$

(B2.31)

where ϕ_{r_n}' and ϕ_{r_n}'' are evaluated by expressions similar to eqn.(B2.29).

Thus, using the necessary boundary condition that relates the calculated magnetic potential, given by eqn.(B2.16b) to that obtained from eqn.(B2.31), at $r = r_2$

$$\phi_2 = \sum_{n=1}^{\infty}\left\{\left(E_n\,r_2^{\,n} + F_n\,r_2^{\,-n}\right)\sin\,n\theta + \left(G_n\,r_2^{\,n} + K_n\,r_2^{\,-n}\right)\cos\,n\theta\right\}_{n\ odd}$$

$$= \sum_{n=1}^{\infty}\left[\phi_{rn}'\,\sin\,n\theta + \phi_{rn}''\,\cos\,n\theta\right]_{n\ odd}$$

giving

$$E_n r^{\,n} + F_n\,r^{\,-n} = \phi_{r_n}'$$

(B2.32a)

$$G_n r^{\,n} + K_n\,r^{\,-n} = \phi_{r_n}''$$

(B2.32b)

The constants A_n, ... , M_n are then evaluated using eqns.(B2.30) and (B2.32) and relations (B2.17), (B2.18) and (B2.19).

Power flow by Poynting theorem and torque equation

As discussed previously, and knowing the radial flux density and peripheral components of flux density in the airgap and arbor regions, the Poynting vector components on the external and inner surfaces are derived as before. Then integrating over the closed surface (see Fig.B2.8), the net power flow is given by

$$P = K_1\left[\int_{S'}\left(\overline{B}_{r_o} \times \overline{H}_{\theta_o}\right)\cdot d\bar{s} + \int_{S''}\left(\overline{B}_{r_i} \times \overline{H}_{\theta_i}\right)\cdot d\bar{s}\right]$$

(B2.33)

where K_1 is a constant.

If P_n and Q_n are the coefficients of $\sin\,n\theta$ and $\cos\,n\theta$ of B_{r_o} and R_n and S_n are the coefficients of $\cos n\theta$ and $\sin\,n\theta$ of H_{θ_o}, derived from eqn.(B2.16a), and similarly if P_n', Q_n', R_n' and S_n' are the corresponding coefficients of B_{r_i} and H_{θ_i}, the final torque expression is obtained as[22]

$$T = 2\pi L \sum_{n=1}^{\infty}\left[r_2^2 \left(P_n S_n + Q_n R_n \right) - r_1^2 \left(P_n' S_n' + Q_n' R_n' \right) \right]_{n \text{ odd}} \text{Nm} \qquad (B2.34)$$

EXPERIMENTAL VERIFICATION

The theory was applied to the same machine, already described for which experimental torque-excitation characteristic was obtained using the torque arm and output meter.

The Constants ϕ_n, ϕ_{r_n}', ϕ_{r_n}''

With the design parameters like rotor external radius, the pole arc and thickness of the annulus known and in accordance with eqn.(B2.27), the potential of the field-system poles is obtained as

$$\phi_p = 1.467 \times 10^{-2}\theta + 1.778 \times 10^{-4}\theta^2 \qquad 0 < \theta < 30° \left(= \alpha'\right)$$

$$\phi_p = 0.2616\,\theta - 1.54 \times 10^{-2}\,\theta^2 + 2.45 \times 10^{-4}\,\theta^3 \qquad 30° < \theta < 37° \qquad (B2.35)$$

where θ is in $°E$ and ϕ_p is in pu ($\phi_m = 1.0$). ϕ_n is then deduced from eqn.(B2.29) and Fourier analysis of rotor-magnetisation mmf yields

$$\phi_{r_n}' = K_2 \sin n\alpha \text{ and } \phi_{r_n}'' = K_2 \cos n\alpha \qquad \text{where } K_2 = 8/n^2\pi^2 \qquad (B2.36)$$

Calculated and Measured Torque-excitation Curves

The developed torque was computed at various 'key' excitations. The highest mmf harmonic order, n, was considered to be 399. The values of computed and measured torques are compared in Table B2.3 showing that the theory described above would predict the torque-excitation characteristic of a salient-pole coupling with sufficient accuracy in the working range.

Table B2.3: Computed and measured torque-excitation characteristics of the experimental coupling

Excitation, A	Torque, Nm	
	computed	measured
0.1	0.001	0.0048
0.15	0.0163	0.015
0.2	0.0355	0.04
0.25	0.0421	0.046
0.3	0.0516	0.048

---

[22]Expression (B2.34) is directly applicable to a coupling having two poles. Where the coupling comprises p pole pairs, the torque will be (theoretically) p times that given by eqn.(B2.34). This follows from the basic property of the developed torque being proportional to total hysteresis loss in the rotor active material or

$$T \propto f V \int H \, dB$$

where f is the cycle of hysteresis loops per revolution and V is the volume of the rotor hard material.

The theory is based on the permanent-magnet properties of the rotor annulus material being represented as a secondary, 'active' member, without any currents. This allows its 'reaction' to be incorporated in the analysis in the form of an independent Laplacian magnetic scalar potential, to predict a quantitative performance of the salient-pole hysteresis coupling. A qualitative performance has been discussed before[23].

[23]See, for example,

M.J.Jevons and S.C. Bhargava: The salient-pole hysteresis coupling, IEEE Trans. Magnetics, Vol. MAG-11, No.5, 1975, pp 1461-63.

3 : The Hysteresis Brake

3

The Hysteresis Brake

MECHANICAL FUNCTION OF A BRAKE

A brake is a mechanical device that when used or activated inhibits motion of a moving body or system by absorbing its energy, stored during motion. It may be used for slowing or stopping a moving vehicle, wheel or axle altogether, or to prevent its motion.

Brake Action

Most brakes commonly use friction between two surfaces pressed together on to the moving body or a part attached to it, to convert the kinetic energy of the moving object into heat, though other methods of energy conversion may be employed; for example, "regenerative braking" that converts much of the energy into electrical energy in electric locomotion.

Types of Brake

Brakes may be broadly divided into two types:

- mechanical, or
- non-mechanical

Mechanical brakes

These may be classified as frictional brakes of

A. DRUM

or

B. DISC

type using friction "pads" or "shoes".

Drum-brake assembly

A drum brake is one in which the friction is effected by a pair of brake shoes that press against the inner surface of a rotating drum. The drum is connected to the rotating wheel hub, say, in a vehicle. Although these were commonly employed in cars and trucks in old times, their use is now mostly confined to motor cycles and other two-wheelers. A serious disadvantage of this type assembly is that brake shoes wear out faster due to their tendency to overheat, requiring frequent maintenance. A typical drum-type brake assembly is shown in Fig.B3.1.

Disc type brake

The disc brake is a device for slowing or stopping the rotation of wheel(s) of a vehicle incorporating a "disc" in place of a drum. The brake disc, usually made of cast iron or ceramic, is connected to the wheel or the axle. To stop the wheel, frictional material in the form of brake pads, mounted in a device called a brake calliper, is activated mechanically[1] to actuate on both sides of the disc, the resulting friction slowing or stopping the vehicle.

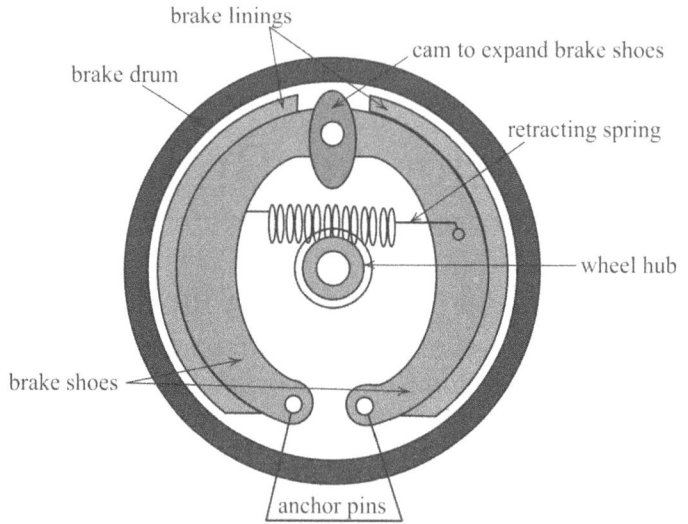

Fig.B3.1 : A drum brake assembly

The details of a disc brake assembly is shown in Fig.B3.2.

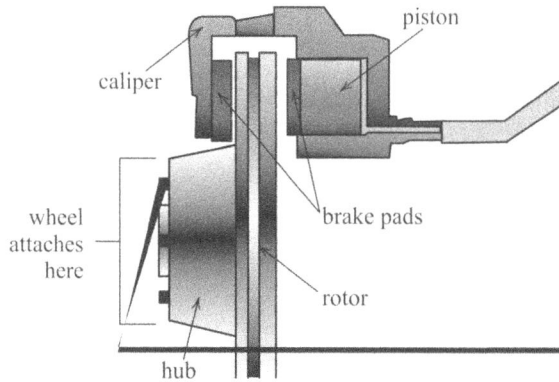

Fig.B3.2 : Details to show working of a disc brake

Electromagnetic Brakes

Electromechanical brakes

The earlier form of electromagnetic brakes in common use would be known as electromechanical brakes comprising an electromagnet acing on an armature to activate

---

[1]In modern cars and other four-wheelers, this is achieved hydraulically or pneumatically.

the brake action on a "friction disc" device. The basic form of one type is illustrated in Fig.B3.3.

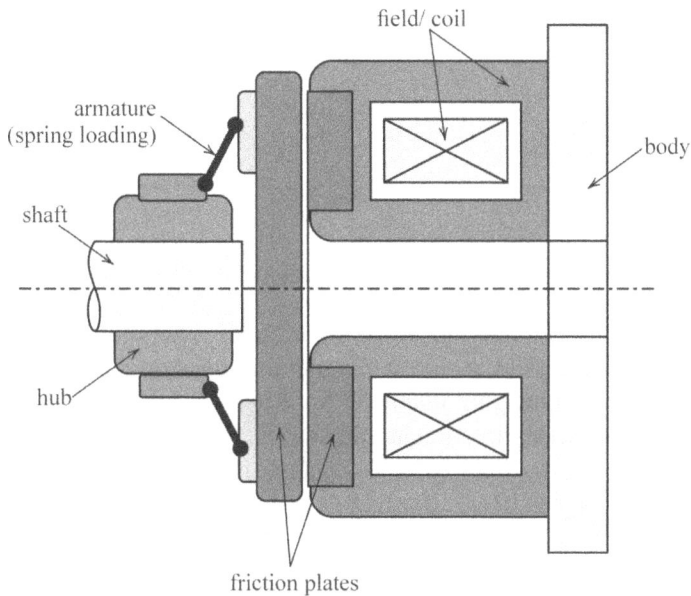

Fig.B3.3 : Assembly of an electromechanical brake

When current is applied to the field/coil, it creates a magnetic field that attracts the armature to the face of the brake. The armature and hub are normally mounted on the shaft that is rotating. Since the brake coil is mounted solidly, the brake armature, hub and shaft come to a stop in a short time. When the current is (switched) off, the armature is free to turn with the shaft. In most designs, springs hold the armature away from the brake surface when power is released, creating a small airgap.

The brake allows for a very fast response, providing a smooth and quiet operation. The modern forms of electromagnetic brakes are classified as

- Eddy current brakes
- Hysteresis brakes

Eddy current brakes

In these type of brakes, the braking action results from the interaction between magnetic field of a permanent or electromagnet and induced eddy currents in a *moving,* electrically conducting metal object, for example a copper disc, based on the principle of electromagnetic induction.

By Lenz's law, the (closed) induced currents create their own magnetic field which opposes the field of the magnet. Thus the moving metallic medium (disc or plate) would experience a drag force or braking action that opposes its motion.

The principle of operation of an eddy current brake is illustrated in Fig.B3.4.

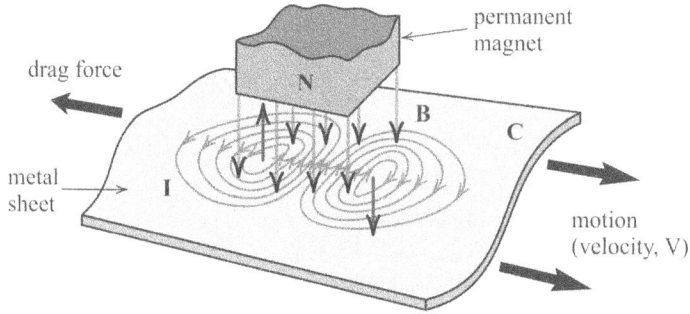

Fig.B3.4 : To show principle of operation of an eddy current brake

It shows a metal sheet, C, moving to the right under a permanent magnet (only north pole shown). The magnetic field, B, of the magnet impinges on the sheet below. Since the metal is moving, the time-changing magnetic field induces eddy currents, I, in the sheet, coplanar and in anticlockwise direction as shown.

The interaction of the currents with the magnetic field results in a force in a direction opposite to the motion of the sheet (indicated by V), resulting in the braking action.

Applications

Eddy current brakes find use in a multitude of applications. Some of these are

- Meg-lev trains
- Industrial machines and tools
- Gym equipment
- Recreation rides and roller coasters
- Conventional induction-type energy meters to provide braking torque

Hysteresis brakes

These operate based on the use of hysteresis loss in hard magnetic materials, incorporated suitably in the brake assembly.

General

Hysteresis brakes are essentially a specially-designed electromagnetic device, similar to most other hysteresis machines, that produce torque strictly within a magnetic airgap, without the use of any friction pads or linings that characterise mechanical brakes over which they have several advantages, having far superior operating characteristics.

These brakes possess extremely wide torque range and can be controlled remotely, being ideally suited for test stand applications where varying torque is required.

Construction

The schematic construction of a hysteresis brake assembly is shown in Fig.B3.5 whilst cut-away section of an actual brake is shown in the photographic view in Fig.B3.6.

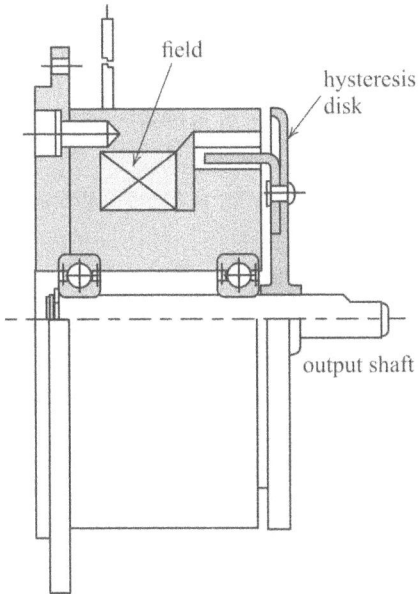

Fig.B3.5 : Schematic of a hysteresis   Fig.B3.6 : Photographic view of a hysteresis brake
           brake assembly

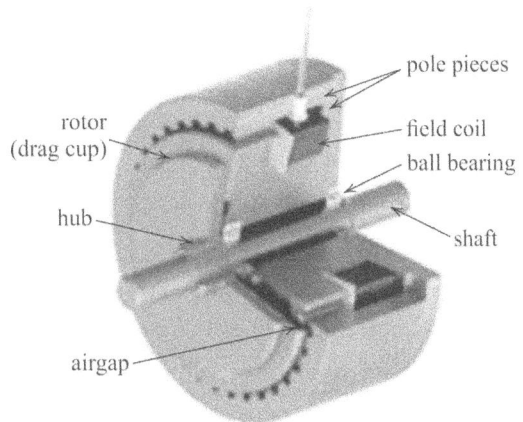

The 'excitation' part of the assembly consists of a field system carrying a three-phase distributed winding, connected to a balanced three-phase supply. When excited, this would produce a rotating magnetic field in the airgap of the assembly. The output shaft of the machine that is required to be slowed down or braked carries a disc made of a hard magnetic material having high hysteresis loss (per cycle and per unit volume), firmly fitted to the shaft and rotating with it in either direction.

Operation

Under the influence of the rotating magnetic field, the high-hysteretic disc develops a torque in a direction opposite to the direction of rotation of the shaft thus resulting in a braking action. Clearly, the magnitude of developed torque would depend on

- the excitation current and corresponding flux density in the airgap
- the volume of the hysteresis disc
- the hysteresis coefficient of the material
- the speed of rotation or corresponding frequency of flux impinging on the disc as in any hysteresis machine[2].

With the last three parameters remaining un-changed, the developed torque can be controlled by varying the excitation.

---

[2]See, for example,

W. C. Orthwein: Clutches and Brakes: Design and Selection (book), CRC Press, 2004 for a detailed discussion about hysteresis brakes.

Advantages

Since the torque in hysteresis brakes is produced without any physical contact of parts, these devices are not subject to wear (except the normal wear of anti-friction bearings). This feature makes them distinctly superior to mechanical-friction brakes in life expectancy, servicing requirements and constancy of performance. Hysteresis brakes are also the most 'repeatable' braking devices known; they will repeat their performance precisely, an indefinite number of times, whenever operating factors are altered.

Hysteresis brakes are also stable in practically any environment. They are not damaged by reasonable temperature cycling, and can operate as hot as oil and bearing lubricants will tolerate. . When operating by fixed current they show no significant torque variation even in extreme ambience. They also have the widest speed range of all electric torque-control devices, from zero to a high speed determined by kinetic power dissipation and the physical size of the unit.

Their power consumption is extremely low. Since their working members have no physical contact and thus can accept moderate expansion without effect on operation, they can be readily adapted for use in high-vacuum applications.

In addition, the simple, non-complex design features of hysteresis brakes provide several inherent advantages over other types of brakes, employing friction pads or discs.

The advantages can be summarised as

- no friction devices and hence no wear and maintenance requirement
- corresponding much longer life
- quiet operation, without vibrations and noise
- superior torque repeatability
- broad speed range
- excellent 'environmental' stability
- easy and smooth control of developed torque

Disadvantages

Hysteresis brakes are characterised by a few inherent disadvantages such as

- the torque output is not linear with the input current
- the developed torque exhibits 'hysteresis' with respect to the applied current; the torque follows a curve with increasing current, but does not follow the same curve as the current is decreased (see Fig.B3.7)[3]

---

[3]This may be explained by reference to hysteresis loop(s) of the hard magnetic material of the disc.

- although often stated, the hysteresis torque is not quite independent of speed
- the developed torque is at times compounded by eddy-current torque which may be difficult to assess

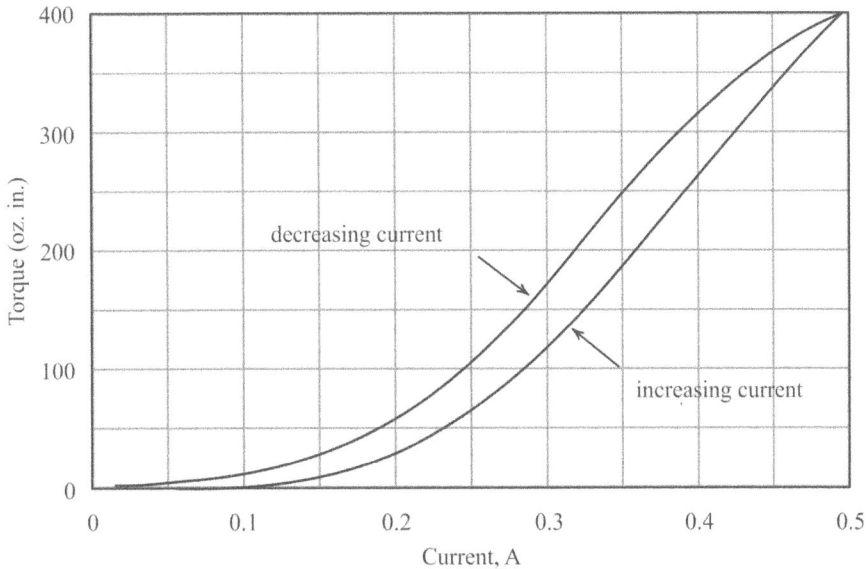

Fig.B3.7 : Developed torque in a hysteresis brake showing hysteresis effect

Applications

Hysteresis brakes provide precise control of tension and find applications in machines and industries such as

- armature and coil winding
- printing and labelling
- wire making
- braiding
- material slitting
- sheeting
- weaving

Some the applications are illustrated in Fig.B3.8.

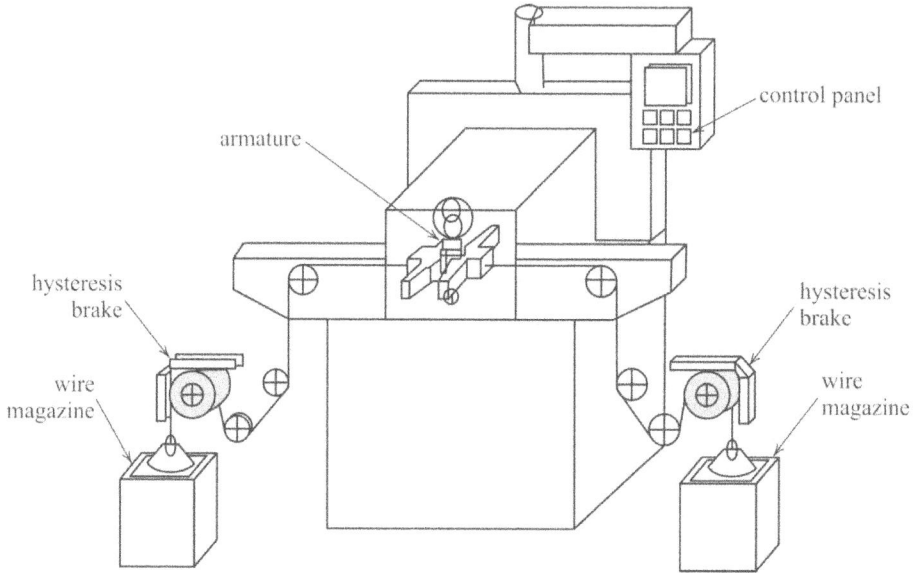

(a) tension control during wind, hook and cut operations of high speed automated winding machines

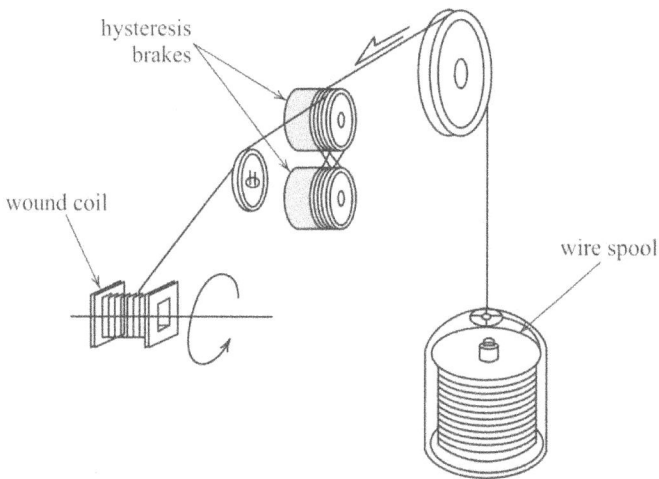

(b) transformer and coil winding operations employing hysteresis brakes in open loop control for maintaining precise tension during winding process

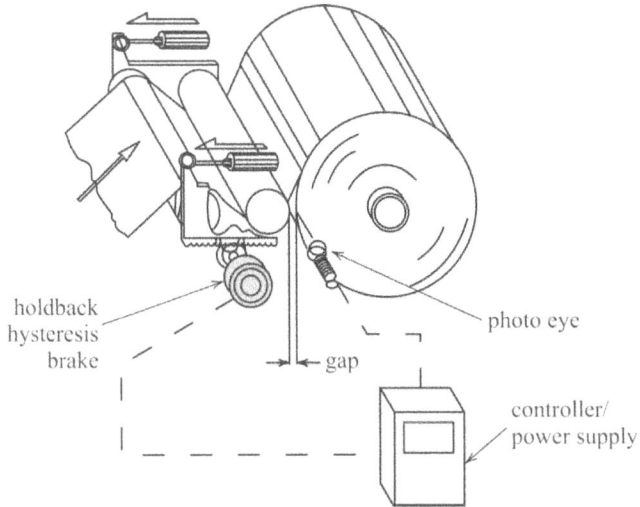

(c) precision gap control using hysteresis brake and photo sensor in paper making

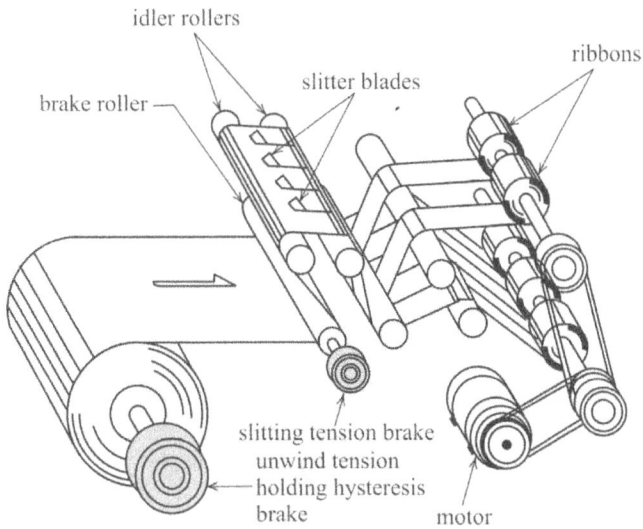

(d) frictionless, non-breakaway force for tension control during slitting operation industry

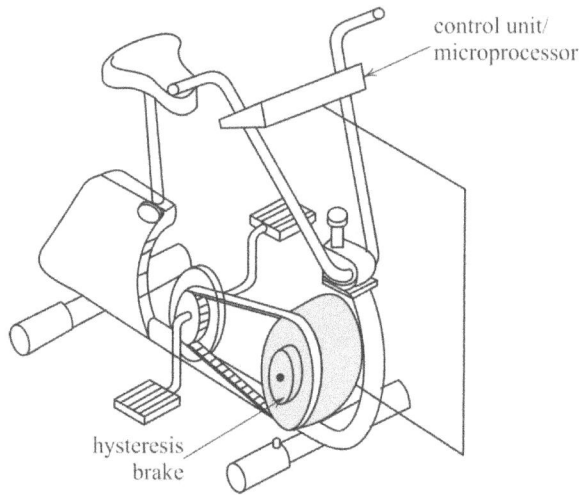

(e) hysteresis brake in use in an exercise bicycle

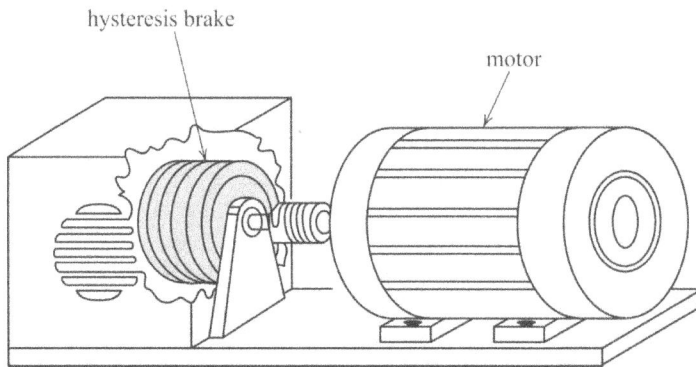

(f) hysteresis brakes used as the preferred method of braking in a dynamometer

Fig.B3.8 : Illustrations of some of the applications of hysteresis brake

The other areas of use of hysteresis brakes include

- medical technology
- packaging machines
- draught regulation with thread-, wire-, cable-, rope-, foil-, paper- film and strip tensions
- printers, copiers
- money- and vending-machines; for example ticket vending machines

Cogging

A somewhat undesirable phenomenon associated with hysteresis brakes is known as "cogging".

This occurs esp. in electromagnetic field system comprising salient poles that leave an imprint of poles of opposite polarity on the hard-magnet disc as a result of residual magnetism when supply to the field system is reduced to zero.

One way to eliminate "cogging" (at least in the case where the primary magnet is an electromagnet) is to first increase the input electrical power to the electromagnet to the highest value previously used and thereby reestablish the highest output torque before cogging occurred.

4 : The Hysteresis Clutch

4

The Hysteresis Clutch

CLUTCH

General

A **clutch** may be described as a mechanical device which engages and disengages power transmission especially from a driving shaft to the driven shaft. Thus, clutches simply connect and disconnect two rotating shafts.

In a typical clutch assembly, one shaft is attached to an engine or other power unit (the driving member) while the other shaft (the driven member) provides output power for work. Whilst usually the motions involved are rotary, linear clutches are also possible.

As a simple application, consider a torque-controlled (electric) drill in which one shaft is driven by a motor and the other drives a drill chuck. The clutch connects the two shafts so they may be 'locked' together and spin at the same speed; the term being used as "engaged", locked together but spinning at different speeds, that is "slipping", or unlocked and spinning at different speeds or "disengaged".

Clearly, a clutch represents a device that may be used to connect the driving shaft to a driven shaft, so that the driven shaft may be started or stopped at will, without stopping the driving shaft. A clutch thus provides an interruptible connection between two rotating shafts, allowing a high inertia load to be started with a small power.

A popularly known and important application of a (mechanical) clutch is in automotive vehicles where it is used to connect the engine and the gear box. Here the clutch enables the vehicle engine to be started before the rotating motion of the engine shaft is transmitted to the wheels through the propeller shaft; while also facilitating changing of gears as required. Clutches may also be used commonly in various production machinery of all types.

A mechanical clutch for use in a car

A "disc" type clutch, typically used in a car, is shown schematically in Fig.B4.1 to demonstrate action of a clutch.

In a car, a flywheel connects to the engine whilst a clutch plate connects to the transmission, a gear box being interposed between the two.

When the driver's (left) foot is off the pedal, the springs push the pressure plate against the clutch disc which in turn presses against the flywheel. This locks the engine to the transmission input or propeller shaft, causing it to spin at the same speed. The propeller shaft then transfers the motion to the rear wheels of the car on either side through the differential.

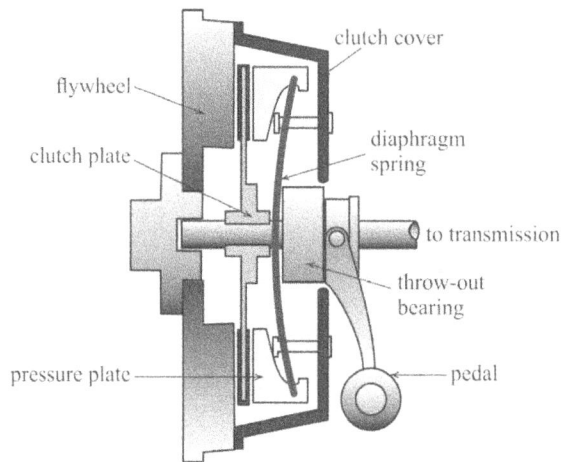

Fig.B4.1 : A disc type clutch used in a car

Pressing the (clutch) pedal, disengages the clutch plate or disc from the fly wheel by the action of the "diaphragm" spring, allowing the driver to change the desired gear in the gear box, followed by gently releasing the clutch pedal to engage the propeller shaft.

The amount of force the clutch can hold, and hence its operational efficiency, depends on the friction between the clutch plate and the flywheel and the force the spring puts on the pressure plate.

Electromagnetic Clutch

An **electromagnetic clutch** offers a better control on the developed torque required for the clutch action as compared to a simple mechanical clutch and hence a more reliable operation.

These clutches operate electrically but transmit torque mechanically. Earlier known as electromechanical clutches, over the years EM came to stand for electromagnetic, referring to the way the units actuate, but their basic operation has not changed.

Electromagnetic clutches are most suitable for remote operation since no mechanical linkages are required to control their engagement or disengagement thus providing fast, smooth operation.

Working of an electromagnetic clutch

An electromagnetic clutch essentially comprises an 'enclosure' housing coils or winding, an armature, rotor and (output) hub as shown in Fig.B4.2. The armature plate is lined with friction coating or carries a friction disc or plate as depicted schematically in Fig.B4.3. The coil is positioned behind the rotor. When the clutch is activated by energising the coils with direct current, it generates a magnetic field thereby magnetising the rotor. The magnetic field crossing the air gap between rotor and armature pulls the armature toward the rotor. The frictional force generated at the contact surface transfer the torque, the engagement time being dependent on the strength of magnetic fields, inertia of the load and, electromagnetically, on the airgap.

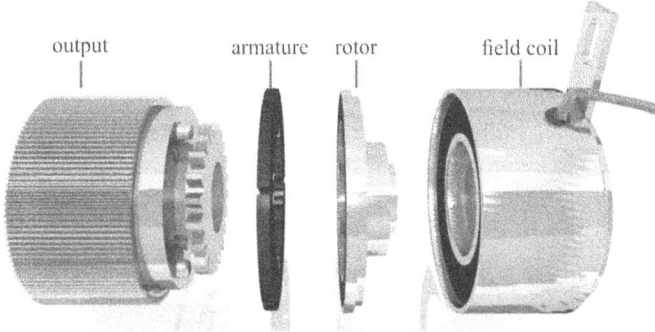

Fig.B4.2 : Components of an electromagnetic clutch

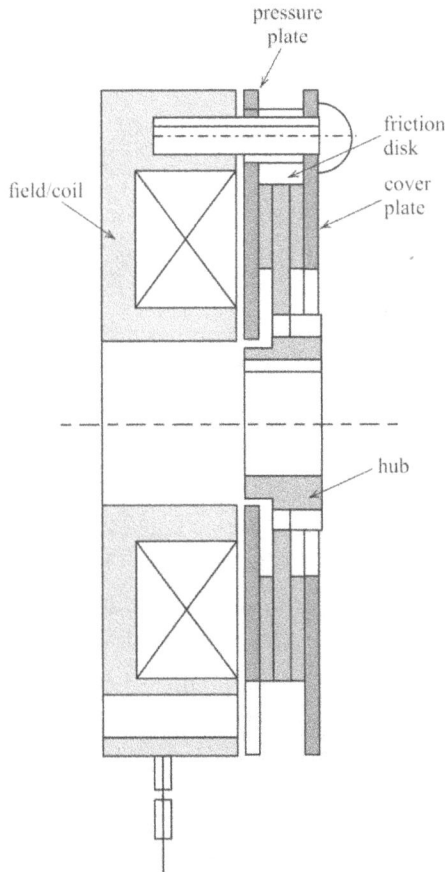

Fig.B4.3 : A single-disc electromagnetic clutch

Single- and Multi-disc Clutch

Single-face or single-disc clutches are extensively employed in numerous applications. However, in case of heavy-duty requirements, multi-disc clutches become unavoidable and are used to deliver extremely high torque in a relatively small space[1].

Operation of a multi-disc EM clutch

The operation of a multi-disc clutch is based on the same principle as a single-disc clutch and can be visualised by reference to the schematic shown in Fig.B4.4.

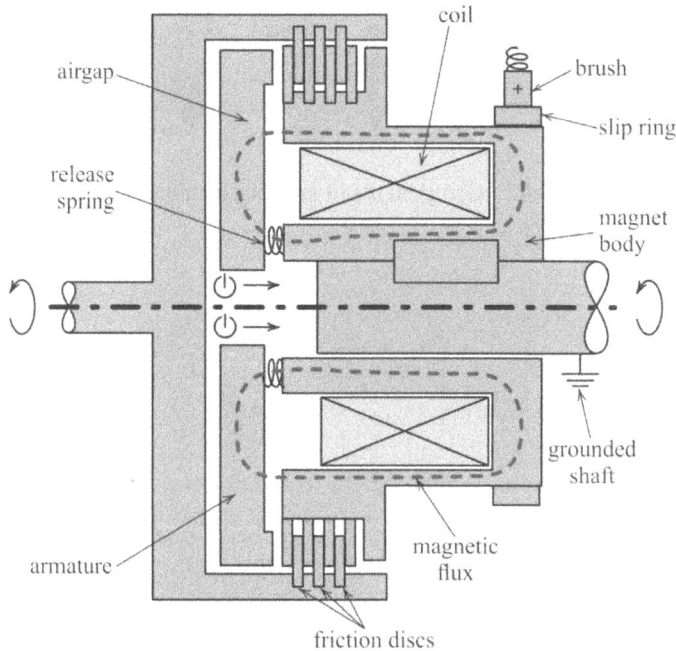

Fig.B4.4 : A multi disc electromagnetic clutch

The attraction of the armature compresses ('squeezes') the friction disks, transferring the torque from the inner driver to the outer discs. The output discs are connected to a gear, coupling, or pulley via a drive cup. The clutch slips until the input and output RPMs are matched. This happens relatively quickly, typically in 0.2 - 2 sec.

When the current is switched off, the armature is free to turn with the shaft. Springs hold the friction disks away from each other, so there is no contact when the clutch is not engaged, creating a minimal amount of drag.

---

[1]These clutches can be used 'dry' or 'wet' (that is in oil bath). Running the clutches in an oil bath greatly increases the heat dissipation capability making the clutch ideally suited for multiple speed gear boxes and machine tool applications.

Technical Features of Electromagnetic Clutches

- Available with various voltage options ('standard' coil voltage being 24 V DC)
- 'Dry' operation
- Available torque range in practice: 1 Nm to 2500 Nm
- Simple design and construction
- High operating efficiency
- Zero backlash
- Horizontal or vertical mounting possible
- Low maintenance
- Remote control is possible
- Fast, smooth operation

Typical Applications

- Electromagnetic clutches have a vast spectrum of applications. Some of the common applications are
- Packaging machines
- Copying and printing machines
- Pharmaceutical machines
- Food processing machinery
- Machine tools (for example an electric drill)
- Conveyers and lawnmowers
- Textile machinery
- Wire-drawing machinery
- Cranes, hoists and miscellaneous material handling equipment
- Factory automation

Two important applications of electromagnetic clutches are in automobiles and locomotives.

Automobiles

When the electromagnetic clutch is used in automobiles, there may be a clutch release switch inside the gear lever. The driver operates the switch by holding the gear lever to change the gear, thus cutting off current to the electromagnet and disengaging the clutch. With this mechanism, there is no need to depress the clutch pedal. Alternatively, the switch may be replaced by a "touch" or "proximity" sensor which senses the presence of the hand near the lever and cuts off the current.

The advantages of using this type of clutch for automobiles are that complicated (mechanical) linkages are not required to actuate the clutch, and the driver needs to apply a considerably reduced force to operate the clutch. The clutch action can be reckoned as a type of "semi-automatic transmission".

Locomotives

Two special design requirements of the clutch for use in a locomotive would be to match

- high rotational speeds, up to 3,000 rpm, and
- high torque, up to 5,000 Nm

encountered in a locomotive.

An operating life of over 50,000 hours, without maintenance, would be anticipated since the clutch may be housed in a closed body with little access to any dust or other environmental pollutants. Also, depending on the location or usage, the clutch may be required to operate reliably in extreme ambient temperatures, ranging from, say, –40 °C to 100 °C.

Advantages and Disadvantages

Advantages

- These clutches do not require complicated linkages such as pedals etc., typical of mechanical clutches

Disadvantages

- High initial cost
- Operating temperature of clutches is limited for at high temperature the insulation of the electromagnet may be damaged; also risk of overheating during engagement

Hysteresis Clutch

Hysteresis clutch is the type of electromagnetic clutch having distinctly advantageous features that has an extremely high torque range. The clutch has no mechanical contact between the rotating parts. It belongs to a class of synchronous magnetic coupling, similar in construction to it[2], and is used for non-contact transmission of mechanical torque from the drive shaft to the output or driven shaft. As a result, it provides an extremely long life and gives superior torque repeatability. These clutches ensure a smooth operation with almost no maintenance.

---

[2]See, for example,

S.C.Bhargava: Theory of hysteresis coupling torque, Electric Machines and Electromechanics, Vol. 5, 1980, pp 391-405.

Basic hysteresis clutch

Construction

A hysteresis clutch essentially consist of

- a system of permanent magnets or electromagnetic field system mounted on the driving shaft
- a member carrying a plate or disc of permanent magnet material, say AlNiCo, mounted on the driven shaft to undergo clutching action as depicted schematically in Fig.B4.5

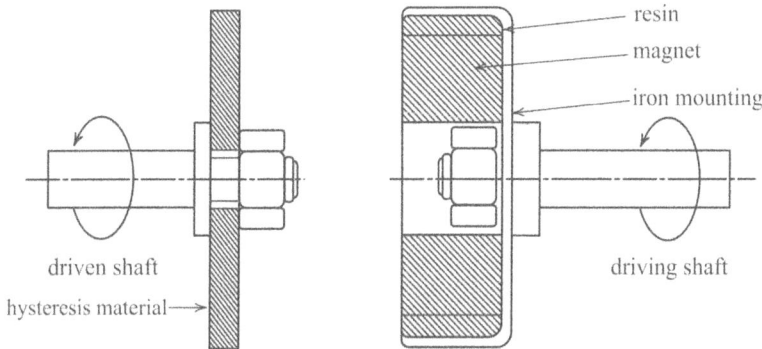

Fig.B4.5 : Schematic of a basic hysteresis clutch

Operation

As the field system is being rotated, connected to a driving motor or device, the mmf of the magnetic field results in magnetisation of the hard magnet following one complete revolution. The disc develops the driving torque corresponding to hysteresis loss in the material, similar to that in a hysteresis coupling or brake, leading to rotation of the shaft to be driven. After a brief time the driven shaft acquires the speed of the driving shaft.

Hysteresis clutch in practice

A cut-away schematic of a hysteresis clutch in practice is shown in Fig.B4.6.

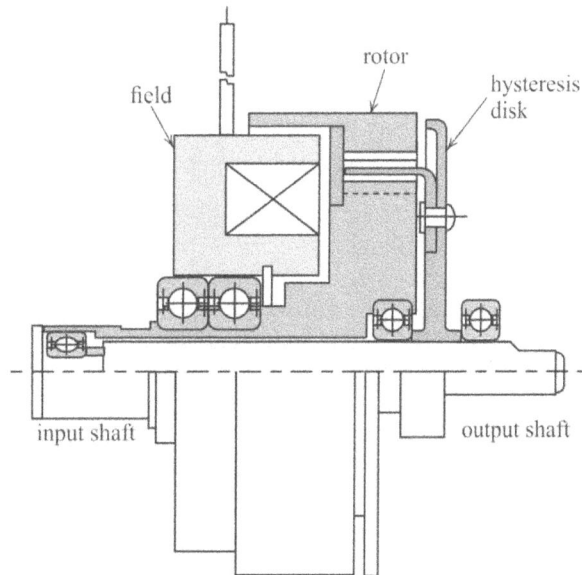

Fig.B4.6 : A practical hysteresis clutch

The clutch comprises an electromagnet having a given number of poles, mounted on the driving shaft and connected to the 'rotor' supported on bearings as shown. The disc made of hysteresis material is mounted on the driven shaft and its 'flat' part passes through the rotor.

When the field system is excited with a given current, it magnetises the rotor with the alternate north and south poles of the field system. This results in the disc undergoing hysteresis and developing a torque corresponding to hysteresis loss. Thus, as the rotor rotates, the torque developed in the disc causes the driven shaft to follow the driving shaft. Depending on the requirement of the output torque, the driven shaft matches the rotation of the driving shaft resulting in 100% lock-up between the two speeds.

As the current to the field system is switched off, the rotor is free to rotate for a while and the disc, and hence the load (shaft), comes to rest.

Circumferential and Radial Flux Clutch

In a disc-type clutch, the best permanent-magnet material would be an anisotropic one. However, an anisotropic material would exhibit its superior performance only along one axis – the preferred axis. This is because nearly all anisotropic hard magnetic materials are "domain-oriented"[3], usually when cooled above their Curie point in a magnetic field directed as desired. Thus, in a circular, ring-shaped disc in the clutch, it is not possible to impart a circumferential preferred magnetisation, resulting in optimum performance of the clutch.

This 'drawback' is circumvented by adopting a "claw" like construction of the clutch in which the hysteresis element or part of it is made of a number of wedge-shaped pieces or blocks of the hard magnetic material, mounted uniformly on a soft magnetic core. This would lead to the preferred axis of magnetisation of *each block* to be radial when magnetised using a suitably designed electromagnet[4].

The path of magnetic flux in such a clutch can be explained by reference to the diagram in Fig.B4.7.

Construction of the clutch

Referring to Fig.B4.7, the hysteresis element consists of a number of 'blocks' of hard magnetic material, M, mounted on a soft magnetic core, R, the preferred direction of magnetic field being radial.

---

[3]See, for example, the classic by Bozorth.
[4]See, for example,
D.G.Young: The hysteresis clutch, Jour. IEE, Vol.9, 1963, pp 437-39.

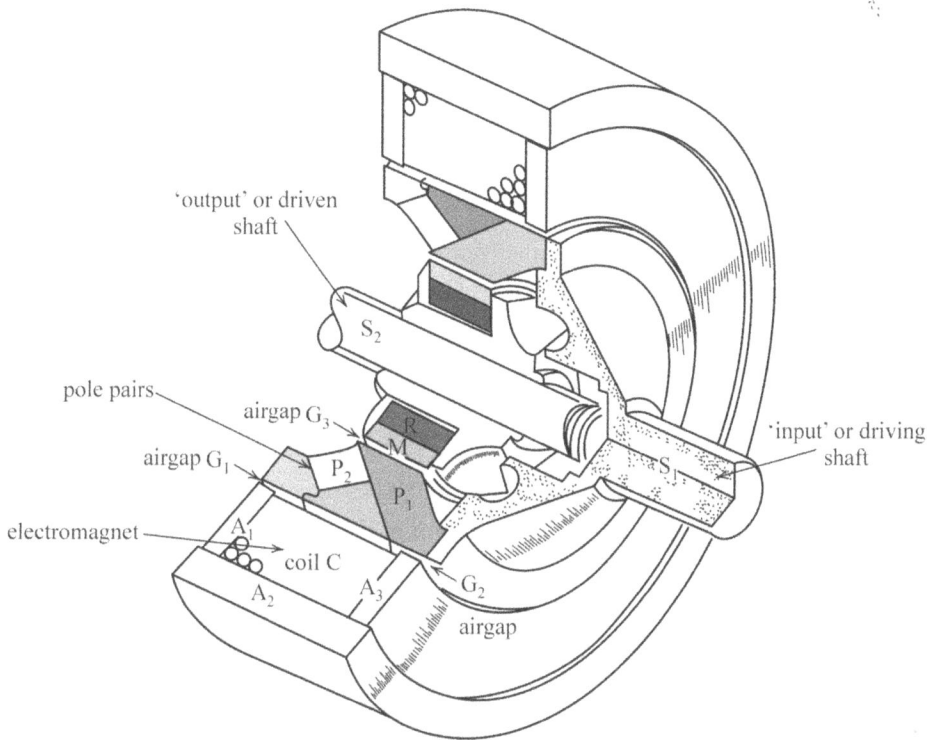

Fig.B4.7 : Magnetic flux in a radial-type clutch

The electromagnet

This comprises

- a set of excitation coils or winding, C, wound suitably and housed in a magnetic case $A_1A_2A_3$, the coil itself being held stationary and excited externally;

- pole pairs P_1P_2, fitted to one of the clutch shafts, connected to the load or driven mechanism, marked S_2.

There are two airgaps G_1 and G_2 between coil and pole-air assemblies and a third gap G_3 between the latter and the permanent-magnet core housing as shown; this constituting the driving member of the clutch, marked S_1.

The magnetic circuit

Disregarding any leakage flux, the path of magnetic flux is from the coil housing A_1A_2 across the outer airgap G_2 into pole P_1, followed by the flow across inner airgap G_3 and then *radially* through the hysteresis 'ring' and finally into the core R. The magnetic path is completed circumferentially around R, radially through M and back through $G_3P_2G_1$ and A_1.

TYPICAL APPLICATIONS

Hysteresis clutches are uniquely suited in various applications where a remote control of variable torque is required.

They are suited where strict precision is required, such as bolting, bottle capping machine, and other 'screw' applications. Their clean nature eliminates the fear of contamination, so they are ideal for food processing system.

Some of these are

A. Closed-loop Winding Tension and Speed Control

In this application, commonly observed in a paper or textile industry, the hysteresis clutch is employed to control the forward movement of the "take-up" drum as the clutch engages and disengages. The tension of the drive is controlled by means of a "drum roller", the electrical control of its movement being obtained by the combination of a potentiometer, the clutch and the controller.

The arrangement is depicted schematically in Fig.B4.8.

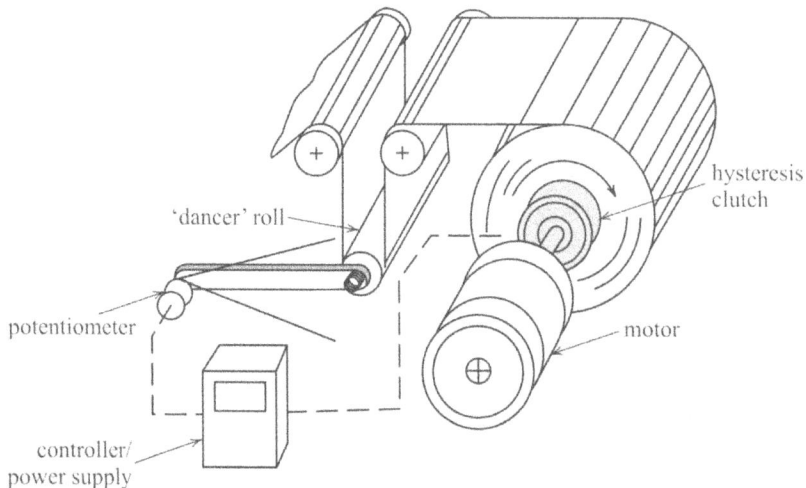

Fig.B4.8 : Closed loop winding and speed control using hysteresis clutch

B. Bottles Capping Industry

The schematic of this process industry is shown in Fig.B4.9.

The key requirements of this industry are

- synchronised movement and operation of the cap-fitting mechanism
- accurate, precise and fool-proof operation of various clutches
- relatively fast operation
- automatic stoppage of the sequence in case of a mal-operation to avoid waste

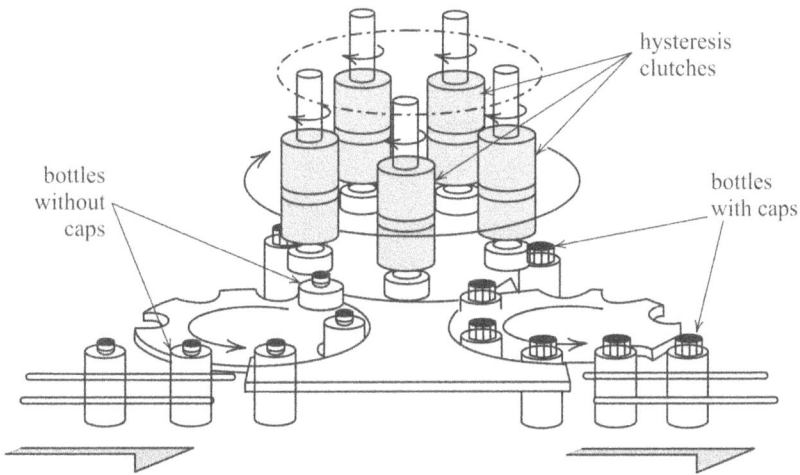

Fig.B4.9 : Use of hysteresis clutches in "capping" industry

As shown, as the bottles without caps move on in a cyclic manner, a spindle with a cap moves down and, with the clutch action, positions the cap on the mouth of the bottle and screws it tight with a precise value of torque; the bottle then moves on fitted with the cap.

Other applications of hysteresis clutches include food-processing industry, for example for capping bottles with juice etc., and clean-room environment on account of their "clean" and smooth nature of working.

5 : The Hysteresis-Reluctance Motor

5

The Hysteresis-Reluctance Motor

RELUCTANCE MOTOR

General

As the name implies, a **reluctance motor** operates based on the phenomenon of magnetic reluctance across the airgap of the machine. The stator of the motor, being of salient-pole construction, is wound to produce a rotating magnetic field in the airgap. The rotor comprises salient ferromagnetic poles consisting of stampings of silicon steel, but *carries no winding*. When excited, the stator field induces (non-permanent) magnetic poles on the rotor protrusions. The action of magnetic reluctance between the two sets of poles produces the desired torque in the motor.

There are four general types of reluctance motors according to their design aspects:

- synchronous reluctance
- variable reluctance
- switched reluctance
- variable reluctance 'stepping'

of which the first two are in common use.

Being of rather robust construction, reluctance motors are capable of delivering very high mechanical power at relatively low cost, making them ideally suited for a variety of applications. Some of the disadvantages are torque ripple, arising from the difference between maximum and minimum torque during one revolution as the rotor 'passes' from maximum to minimum reluctance position when operated at low speed, resulting in consequent noise.

Construction

The basic construction of a '2-pole' reluctance motor is shown in Fig.B5.1(a) whilst the schematic of a motor with six stator and four rotor poles is depicted in Fig.B5.1(b). Notice the concentrated windings on the stator poles.

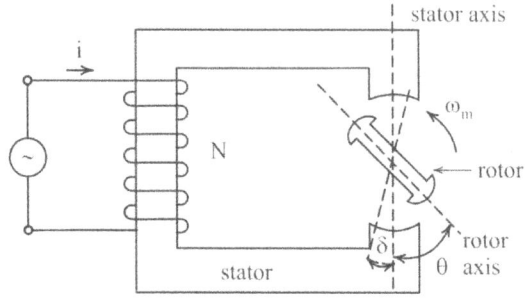

(a)

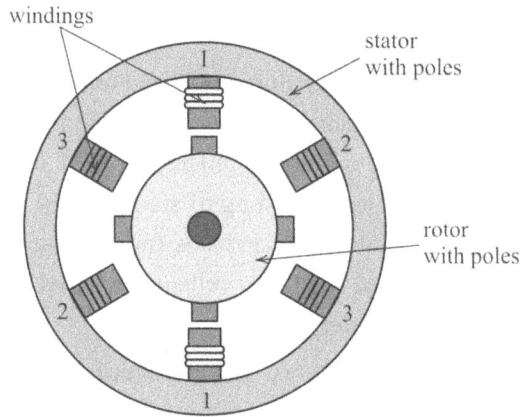

(b)

Fig.B5.1 : Constructional features of a reluctance motor

A detailed constructional view of a reluctance motor is shown in Fig.B5.2.

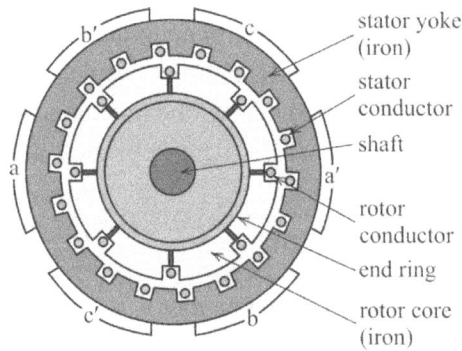

Fig.B5.2 : Construction of a typical reluctance motor

Operating Fundamentals

In a reluctance motor, in general and a switched-reluctance motor in particular, the number of rotor poles is typically *less* than those on the stator as in Fig.B5.1(b). This tends to minimise torque ripple and prevents the two sets of poles from aligning simultaneously – or 'locking' with each other – thus leading to production of no useful torque.

When any of the rotor poles is equidistant from the two adjacent stator poles, the rotor pole is in the "least or minimum" aligned position; this also being the condition of maximum reluctance for the rotor pole – a kind of unstable magnetic equilibrium. In contrast, in the fully aligned condition, two (or more) rotor poles are most aligned with two (or more) stator poles such that the former completely face the latter, corresponding to a condition of minimum reluctance for the poles in question.

Assuming one of the stator poles having been energised, with the nearest rotor pole being induced with opposite polarity, the condition corresponds to there being developed a torque in the rotor in the direction that would tend to reduce reluctance between the stator and the rotor. This results in the nearest rotor pole being pulled from its unaligned position into alignment with the stator field; this also representing the position of least reluctance – similar to the phenomenon of pull of the plunger into a solenoid. To sustain a continuous rotation of the rotor, the stator field must advance with respect to the rotor poles, thereby constantly 'pulling' the rotor along in a given direction. In synchronous reluctance motors, this is achieved by providing stator poles carrying a distributed winding, excited by a three-phase supply, similar to that in a three-phase induction motor. However, in a "switched-reluctance" type of motor, the stator pole windings are excited as an action of electronic 'commutation'. This also provides the motor with the advantages of starting as deemed, smooth operation having low torque ripple and effective speed control.

Constant Speed Operation

Very often, reluctance motors are designed for constant-speed operation when fed from a constant-frequency supply into the three-phase stator winding. A 'damper' winding may be fitted circumventing the rotor poles in which the rotating stator field would induce currents, similar to a squirrel-cage induction motor, so that the machine can self-start. After the rotor pulls into synchronism with the stator rotating field, the motor would continue to operate at constant, synchronous speed.

Switched Reluctance Motors

The switched reluctance motor, or SRM, is a form of "stepper" motor, characterised by the use of fewer poles. The phase windings on the poles are electrically isolated from each other, resulting in higher 'fault tolerance' as compared to, say, an inverter-fed induction motor for controlled speed operation. The robust, sturdy construction of the motor results in relatively low cost. Common uses for an SRM include applications where the rotor is to be held stationary for long durations, and in potentially 'explosive' environments such as mining.

Analysis of Operation

An operational analysis of a reluctance motor can be derived as follows.

Consider the elementary reluctance motor shown in Fig.B5.1(a).

Let

- the variation of the inductance of the windings is sinusoidal with respect to rotor position.
- the variation of the inductance with respect to the stator axis is of double frequency and is given by,

$$L(\theta) = L'' + L' \cos 2\theta \qquad (B5.1)$$

- The stator winding is excited by AC supply, such that

$$i = I_m \sin \omega t$$

Then, the energy stored is a function of inductance and is given by

$$W = \frac{1}{2} L(\theta) \, i^2 \qquad (B5.2)$$

The corresponding flux linkage is given by

$$\lambda(\theta) = L(\theta)i$$

Then the torque is given by

$$\begin{aligned}
T &= \frac{\partial W}{\partial \theta} - i \frac{\partial \lambda}{\partial \theta} \\
&= -\frac{1}{2} i^2 \frac{\partial L}{\partial \theta} + i^2 \frac{\partial L}{\partial \theta} \\
&= -\frac{1}{2} i^2 \frac{\partial L}{\partial \theta}
\end{aligned}$$

Substituting the values of i and L

$$T = - I_m^2 \, L' \, \sin 2\theta \, \sin 2\omega t \qquad (B5.3)$$

where ω represents the angular velocity of the rotating rotor.

Finally, the torque equation can be expressed in terms of ω and ω_m as

$$T = -\frac{1}{2} I_m^2 L' \left\{ \sin 2(\omega_m t - \delta) - \frac{1}{2} \Big[\sin 2(\omega_m t + \omega t - \delta) + \sin 2(\omega_m t - \omega t - \delta) \Big] \right\}$$

where $\theta = \omega_m t - \delta$

and δ = rotor position at t = 0 $\qquad (B5.4)$

The above equation gives the instantaneous torque produced in the machine. The average torque would be zero as average of each term in the above equation is zero.

The value of net torque is not zero when $\omega = \omega_m$ and at this condition the magnitude of the average torque is given by

$$T_{av} = \frac{1}{4} I_m^2 L' \sin 2\delta \qquad\qquad (B5.5)$$

In the general torque expression, ω_m represents the synchronous speed and δ is the usual "torque angle". It is seen that the maximum torque would occur at $\delta = 45°$ which is also termed as the "pull-out" torque at synchronism.

Torque-speed Characteristic

A general form of speed-torque curve of a typical reluctance motor is shown in Fig.B5.3. It is to be noted that the starting torque would be dependent on the rotor position, or rather the position of rotor poles, in the motor.

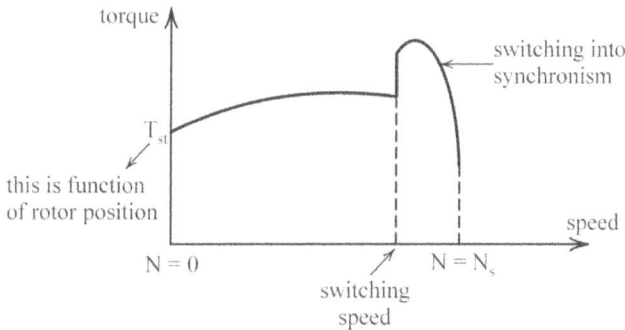

Fig.B5.3 : Torque-speed characteristic of a reluctance motor

HYSTERESIS-RELUCTANCE MOTOR

General

The characteristic feature of a **hysteresis-reluctance** motor is that the developed or shaft torque is the sum total of both: that due to hysteresis action and also due to reluctance action in the rotor[1].

Whilst the hysteresis torque results in self-starting of the machine, providing quiet, synchronous running with constant torque from start to full load, a reluctance motor offers a higher pull-out torque.

Construction

The construction of a hysteresis-reluctance motor can be regarded in terms of a modification of a *circumferential-flux* hysteresis motor with a non-magnetic arbor to incorporate the constructional features of the rotor of a reluctance motor. The schematic

---

[1]It may be noted that, similar to a 'normal' hysteresis motor, there may also be present a variable magnitude of torque developed due to induced eddy currents in the rotor during sub-synchronous operation, thus altering the starting torque of the motor.

cross-section of such an arrangement for a four-pole design is shown in Fig.B5.4; the airgap being shown exaggerated for clarity.

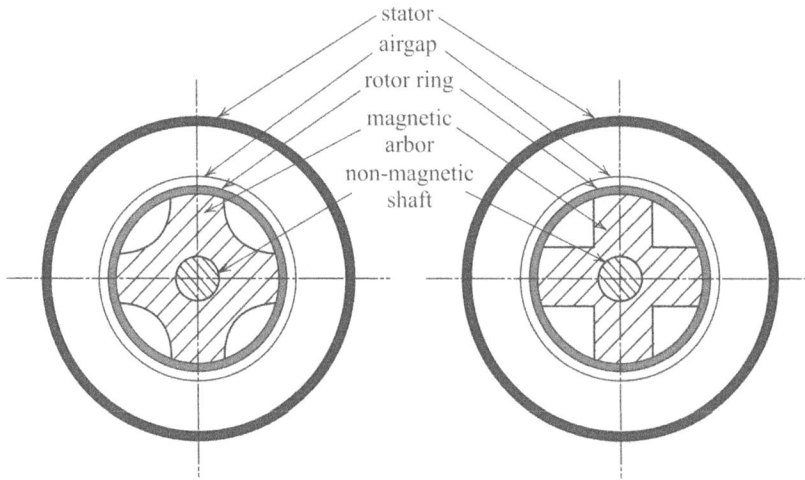

Fig.B5.4 : Cross section of a 4-pole hysteresis reluctance motor

The usual non-magnetic arbor (of the hysteresis motor) is replaced by a (laminated, soft- iron) magnetic spider- or shuttle-shaped support, with projections as shown, to support the hysteresis ring made of a suitable hard magnetic material as usual, mounted on a stainless steel shaft.

Thus, whereas the hysteresis torque is still developed in the same rotor volume as originally in the hysteresis motor, there is now an added reluctance torque brought about by the arbor or rotor poles, resulting in much enhanced pull-out torque.

The stator is similar to that of the usual hysteresis machine comprising a distributed, three-phase winding, wound by adopting the "back-winding" technique[2], and having a 'small' airgap.

Theoretical Aspects

A comprehensive analysis of operational aspects of a hysteresis-reluctance motor may be difficult to obtain, although some attempts have been reported in the past[3].

---

[2]See, for example,

H.C.Rotors: The Hysteresis Motor – Advances which permit economical fractional horsepower ratings, Trans. AIEE, Vol.66, 1947, pp 1419-1430.

[3]See, for example,

M.A.Rahman and A.M.Osheiba: Steady-state performance analysis of poly-phase hysteresis-reluctance motors, IEEE Trans. on Industry Applications, Vol.IA-21, (3), May 1985, pp 659-663.

and

B.J.Chalmers and I.R.Ciric: Performance analysis of hysteresis-reluctance motors with segmental rotors, Proc. IEE, Vol. 121, (9), Sept. 1974, pp 991-992.

As a simple theoretical approach, the combined developed torque of a hysteresis-reluctance motor can be obtained by considering the two effects independently, that is, torque due to hysteresis being developed in the rotor ring and reluctance torque manifesting by action of the arbor poles, and assuming a superposition of the two. A two-pole model is used here to derive the torque expressions; it being assumed that for a multi-pole construction, the final torque expression would be a multiple of number of pole-pairs.

Assumptions

The derivation of the torque expression is based on the following assumptions

- The three-phase stator winding is sinusoidally distributed round the stator periphery and winding or space mmf harmonics are disregarded. Also, the stator is fed from a balanced three-phase supply.

- The (rotor) hysteresis ring (Fig.B5.4) is of 'small' radial thickness, having only a uniform, peripheral magnetisation due to airgap mmf and hence only alternating hysteresis loss.

- The effect of eddy currents induced anywhere in the rotor metallic parts is neglected.

- The iron parts of the machine, except the rotor ring, have infinite rotor relative permeability.

The model

The 2-pole model used for deriving torque expressions is shown in Fig.B5.5.

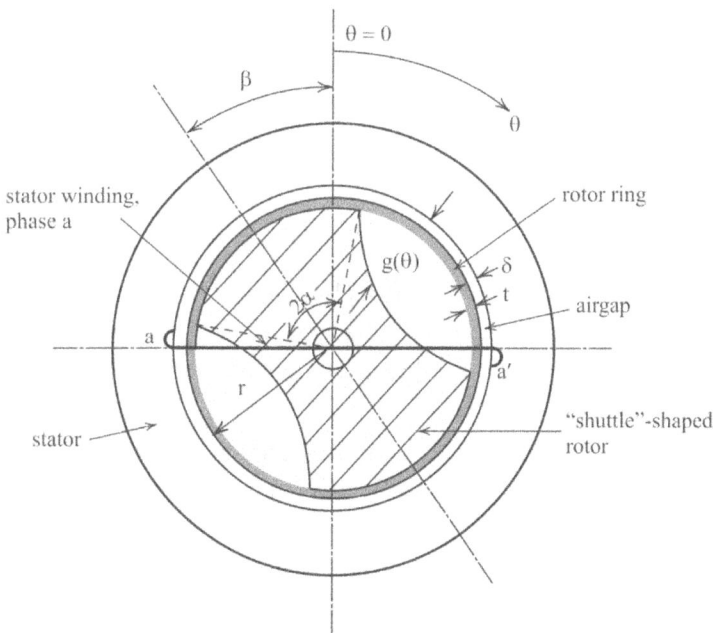

Fig.B5.5 : Model for derivation of torque expression

The torque expressions

A.   Reluctance Torque

This is expressed by

$$T_{rel} = - K \, h_2 \, I_m^2 \, \sin 2\beta_o \tag{B5.6}$$

where (the machine) constant K for a three-phase design is given by

$$K = \frac{3^2 \pi \, I \, r \, \mu_o \, N_s^2}{16}$$

in terms of usual notation.

The details of deriving the torque expression are given in Appendix VII.

Or, in general, for a m-phase machine

$$K = \frac{m^2 \pi \, I \, r \, \mu_o \, N_s^2}{16}$$

With the chosen time reference for the currents in the stator winding, the torque would be zero when β_o equals 0 or $\pi/2$, the condition corresponding to unstable operation from the point of minimum energy requirement. See also eqn.(B5.5) derived earlier. The torque would be maximum when β_o is $\pi/4$. The negative sign indicates that the torque, or the 'equivalent' tangential force at the airgap radius, acts to reduce the angle β_o or tend to align the two magnetic axes.

It can be seen that the torque expression (B5.6) would not be altered if the rotor axis is leading the armature (that is the stator) mmf axis by angle β, instead of lagging behind it as in Fig.B5.5

B.   Hysteresis Torque

This is derived to be

$$T_{hys} = \frac{W_h \, V_{ring}}{2\pi} \quad Nm \tag{B5.7}$$

where W_h represents the alternating hysteresis loss in the hysteresis ring in joule/per unit volume/rev, obtained, for example, from the area of the hysteresis loop for the ring material and V_{ring} is the volume of the ring in m^3, and is seen to be independent of speed as applicable for a hysteresis motor..

Note that the actual torque available at the shaft for a given excitation may be less than the theoretical value on account of the effect of various space mmf harmonics in the machine.

Computation of total developed torque in a two-pole hysteresis-reluctance motor of assumed design parameters is given in Appendix VIII.

Even though researched at length, and having noticeable advantages, hysteresis-reluctance motors seem to have had no commercial exploitation.

Total developed torque

Using principle of superposition, the total developed torque in the hysteresis-reluctance motor would be

$$T = T_{rel} + T_{hys} \ \text{Nm} \tag{B5.8}$$

and for a machine with p pole pairs

$$T = p \times (T_{rel} + T_{hys}) \, \text{Nm} \tag{B5.8a}$$

Part C

EXPERIMENTAL HYSTERESIS MACHINE AND ANALYSIS

1 : General Aspects

1

General Aspects

Genesis of the Experimental Machine

In nearly all conventional hysteresis machines, the rotating magnetic field is produced by three-phase currents fed to a distributed winding in the stator. This invariably gives rise to various space mmf harmonics in the airgap of the machine in addition to the fundamental. The harmonics rotate relative to the fundamental component at speeds inversely proportional to their order. This would introduce difficulty in interpreting the induced EMF waveforms which will not be stationary when viewed, for example, on a CRO, being continuously modified by the harmonics.

When viewed in the rotor reference frame, comprising rotor annulus of a permanent magnet material in a particular form to produce torque in the machine by virtue of hysteresis loss, it would be near impossible to correctly relate the applied mmf to developed torque in the rotor.

Also, a conventional machine or a motor would not suit the experimental research on account of

(a) its rather small mechanical size, in general
(b) compact multi-pole design (for commercial reasons)
(c) possibility of development of eddy-current loss in the rotor caused by fundamental and space harmonics that might be appreciable compared to hysteresis loss and torque esp. if the rotor material is of low resistivity.

Factors (a) and (b) would result in designing a suitable instrumentation in the machine for experimentation to be extremely difficult.

The above aspects led to the design and manufacture of an 'idealised' experimental hysteresis machine[1].

The Idealised Machine

The unique, characteristic feature of the experimental hysteresis machine design was the use of a *mechanically* rotated DC mmf in which the space harmonics would be "locked" to the fundamental component with there being no relative rotation of the various harmonics at all. This permitted the analysis of the applied mmf and resulting torque produced in the rotor considering the fundamental and each harmonic independently, followed by their algebraic superposition to arrive at the resultant developed torque in the machine.

---

[1]In this respect the machine resembled a special form of hysteresis coupling as brought out later.

The rotating field system

The field system comprised two salient poles[2] firmly fitted diametrically opposite to a mild steel yoke, each pole wound with 2200 turns using 22 SWG enamelled insulated wire. The structure resembled field system of a typical 2-pole DC machine[3]. A to-scale plan of the field system showing the arrangement of poles, yoke and excitation winding is shown in Fig.C1.1.

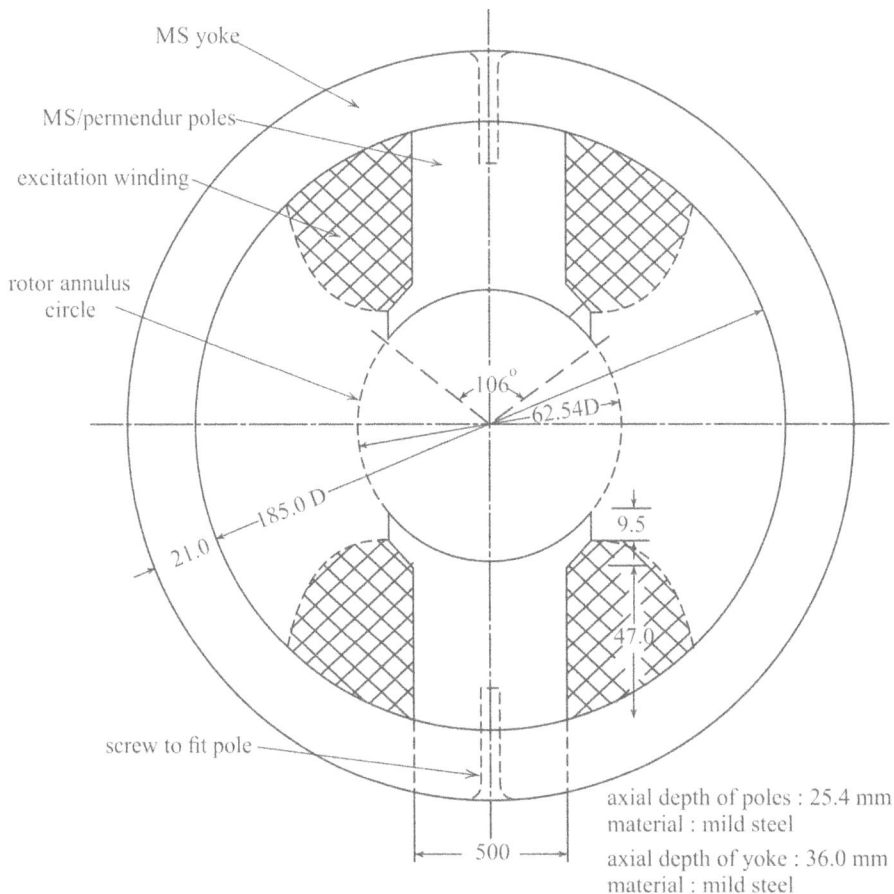

MS yoke

MS/permendur poles

excitation winding

rotor annulus circle

106°

62.54D

185.0 D

21.0

9.5

47.0

screw to fit pole

500

axial depth of poles : 25.4 mm
material : mild steel

axial depth of yoke : 36.0 mm
material : mild steel

Fig.C1.1 : Arrangement of the salient pole field system
(to scale)
dimensions in mm

[2]Of permendur (magnetic material) to avoid saturation or mild steel.

[3]Whilst the field system (or the electromagnet) was rotated mechanically, the rotor was held stationary, or restrained from rotation, resulting in all points of rotor being cyclically magnetised or undergoing hysteresis.

A cross section of the rotatable assembly of the field system is given in Fig.C1.2 whilst Fig.C1.3 illustrates a photographic view of the assembly giving relevant details[4].

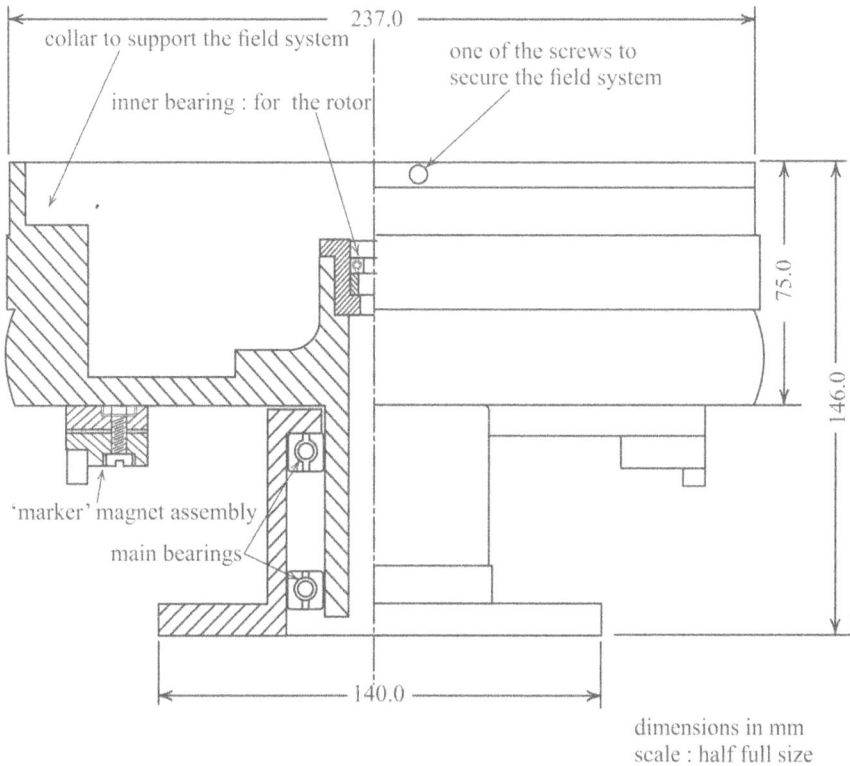

Fig.C1.2 : Cross section of field system assembly

Drive Arrangement

The drive arrangement for the field system was conceived to provide its smooth rotation at a *constant* speed during a test run and also allow the speed to be varied gradually and smoothly from a high value down to standstill when so required.

During the tests, a constant speed of about one rpm was desirable to keep the effect of eddy currents and machine vibrations to a minimum. After each set of tests, typically consisting of gradually increasing the excitation to the field system and measuring the developed torque, the speed was increased by about 2 ½ times and then, together with the excitation current, gradually reduced to zero to demagnetise the rotor.

---

[4]The assembly comprised an aluminium housing, supported on an aluminium pedestal by means of two ball journal bearings, the pedestal being firmly bolted to a bed-plate and the housing being free to rotate. Aluminium was used to keep the weight to a minimum and eliminate leakage flux.

Fig.C1.3 : Photograph of the actual assembly

These requirements were met by using a Velodyne drive motor with a split-field closed loop control system, coupled by a belt to the rotatable housing into which was fitted the field system (see Fig.C1.2). To improve voltage feedback used for speed control, a tachogenerator, driven directly by the housing was used instead of the usual output from the Velodyne tachogenerator. The schematic of the arrangement is shown in Fig.C1.4.

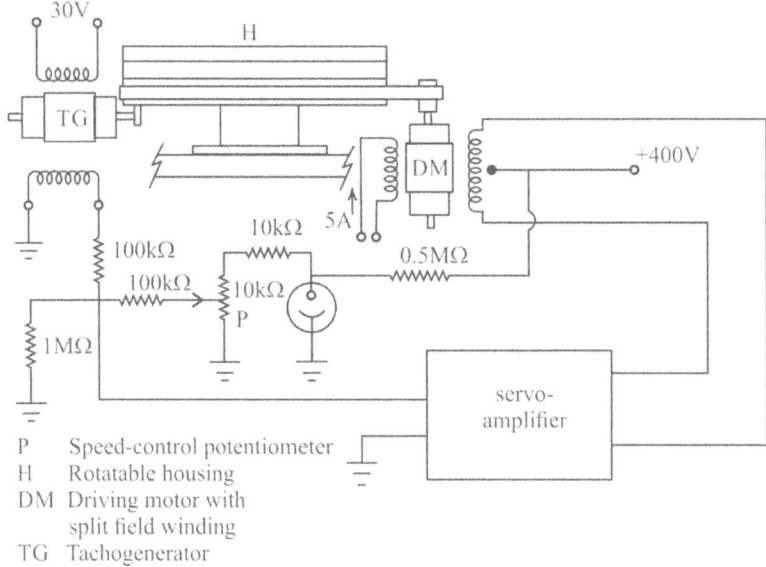

P   Speed-control potentiometer
H   Rotatable housing
DM   Driving motor with
     split field winding
TG   Tachogenerator

Fig.C1.4 : Drive arrangement for the field system

Constant Current Supply for the Field System

It was essential that *excitation current* to the field system, not just the DC supply, remained constant at the pre-set value and not drift during the test on account of, say, change in field winding resistance since any attempt to reset the current would result in a recoil loop in the rotor magnetisation

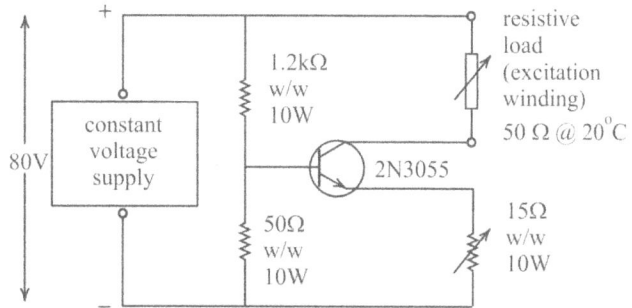

Fig.C1.5 : Constant current supply used during tests

The scheme that converted a stabilised voltage source into a constant-current supply is given in Fig.C1.5 that maintained the highest excitation current of 2.0 A used during tests to within 1 % for a duration of about 5 minutes.

Design of the rotor

The main criterion for design of rotor for the machine was to attain very near *peripheral* flux density distribution within the annulus[5]. This was desirable so that each point on the rotor would undergo only alternating magnetisation cycle so that the assessment of developed torque could be related essentially to alternating hysteresis loss[6]. See Appendix I.

ROTOR CONSTRUCTION

Rotor Annulus

The main constructional details of the rotor are given in Fig.C1.6. The active part of the rotor comprised an annulus made from a vicalloy strip of 25 mm width and 0.4 mm in thickness (as fecilitated by availability), bent in the form of an annulus. A set of very fine holes of 0.2 mm dia were drilled in the strip, prior to forming, for providing search coils as discussed later in detail. The two ends were finished to a close butt joint[7]. The annulus was then appropriately heat treated.

---

[5]This was assured by the use of a thin vicalloy strip for the annulus, being easily available in sheet form.

[6]In a commercial machine having an outside rotor diameter of 82 mm (as in the experimental machine), the radial thickness of the active (or hysteretic) material may be typically about 5 mm resulting in appreciable radial component of flux density in addition to the peripheral component.

[7]Joining the ends using welding or a similar process was disregarded since this might seriously affect the local magnetic properties of the alloy; a butt joint formed a better choice owing to low relative permeability of vicalloy.

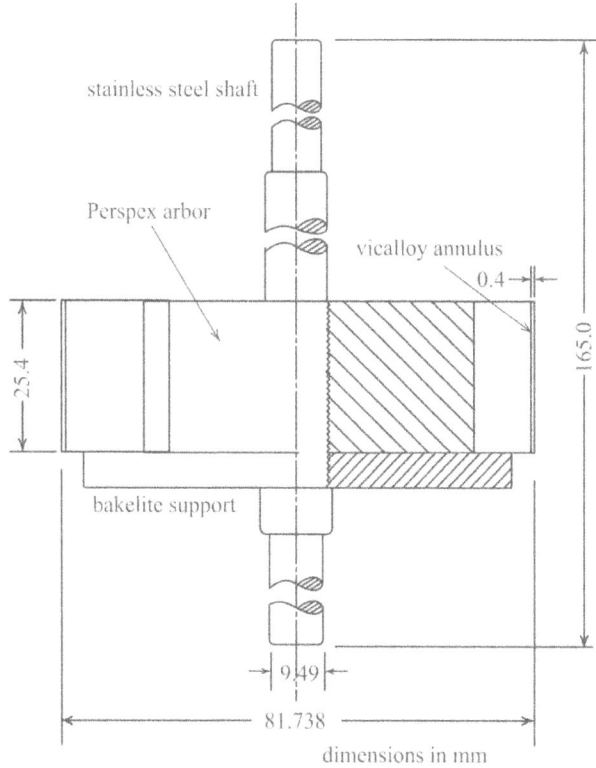

Fig.C1.6 : Constructional details of the rotor

Rotor Arbor

The rotor arbor, the part to support the annulus, was made of Perspex to keep the rotor weight to a minimum causing negligible thrust on the lower bearing of the housing and also to avoid any induced eddy currents. Concentricity of the external surface of the arbor was ensured by its precision machining, followed by making the annulus a close fit on it. Prior to final assembly, three axial slots were also machined to house ('inner') search coils. See Fig.C1.9 later.

Torque Measurement

From mechanical point of view, a precision dynamometer could be used for torque measurement. However, the rotary motion of the dynamometer would result in the rotor to oscillate with respect to the rotating magnetic field and hence modify the magnetisation cycle of the annulus material. Also, the range of developed torque up to maximum excitation would be too low.

This called for an alternative method that would provide precision, being 'stationary' in operation to measure the torque.

Torque arm

The method consisted of using a stainless steel[8] bar of rectangular cross-section, about 30 cm in length. One end of the bar (identified as "torque arm") was fixed to the rotor shaft using an aluminium clamp, the other being held in another rigid clamp (to minimise the effect of vibrations on the measured torque) having a V-shaped notch, thus restraining the rotor from rotating.

The principle of torque measurement was to strain the torque arm by the developed torque and relate the strain to torque in Nm using a suitable transducer. The transducer comprised foil type, standard resistance strain gauges, fixed on the opposite surfaces of the torque arm about 50 mm from the shaft end. Four identical gauges were used in pairs on opposite surfaces of the bar, electrically connected as a bridge.

Calibration of the torque arm

The strain-gauge bridge on the torque arm was supplied from a 1kHz oscillator housed in a commercial "transducer meter". The out-of-balance signal, consequent to production of strain, was amplified within the meter and after demodulation and rectification was measured on a moving coil instrument.

To calibrate the torque arm, the rotor was held between the centres of a lathe, with the arm inclined and its free end resting on a flat support. A non-elastic string was wound round the rotor, with its one end fixed and the other carrying a scale-pan to be loaded with gram weights to strain the arm, the deflection on the transducer meter being noted for different weights. The calibration graph thus obtained is shown in Fig.C1.7.

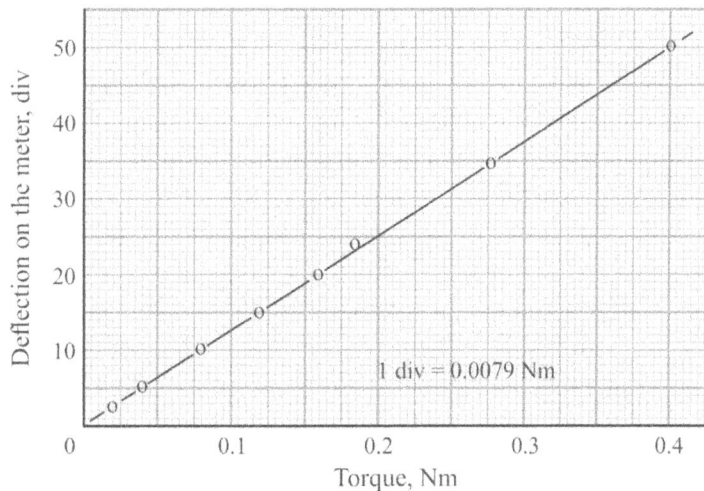

Fig.C1.7 : Calibration curve for the torque arm

Requirement of Concentricity

With the rotor annulus being of low relative permeability and the mean airgap between the annulus external surface pole surface of the field system being only 0.4 mm to keep

---

[8]Preferred to avoid mechanical hysteresis.

the reluctance to a minimum, it was vital to maintain 'absolute' concentricity of the axis of the stationary rotor with the axis of rotation of the field system.

This was achieved by first fitting the rotor vertically between the top and bottom bearings and then aligning the field system in the housing by adjusting three screws successively, fitted equilaterally in the housing. By repeated adjustments, it was possible to adjust the airgap to within ± 0.0254 mm ($\pm 0.001''$)[9]

Measurement of Speed

Accurate measurement of rotation of the field system was essential since it figured in the computation of EMF induced in the search coil(s) and hence flux density in the annulus. An electronic scheme of speed measurement was evolved using a small permanent magnet fitted suitably on the external side of the housing and making use of two stationary reed switches, one of these producing a "start" pulse to trigger an electronic timer and the other resulting in a "stop" pulse to stop the timer after half revolution of the field system. The schematic of the arrangement is shown in Fig.C1.8.

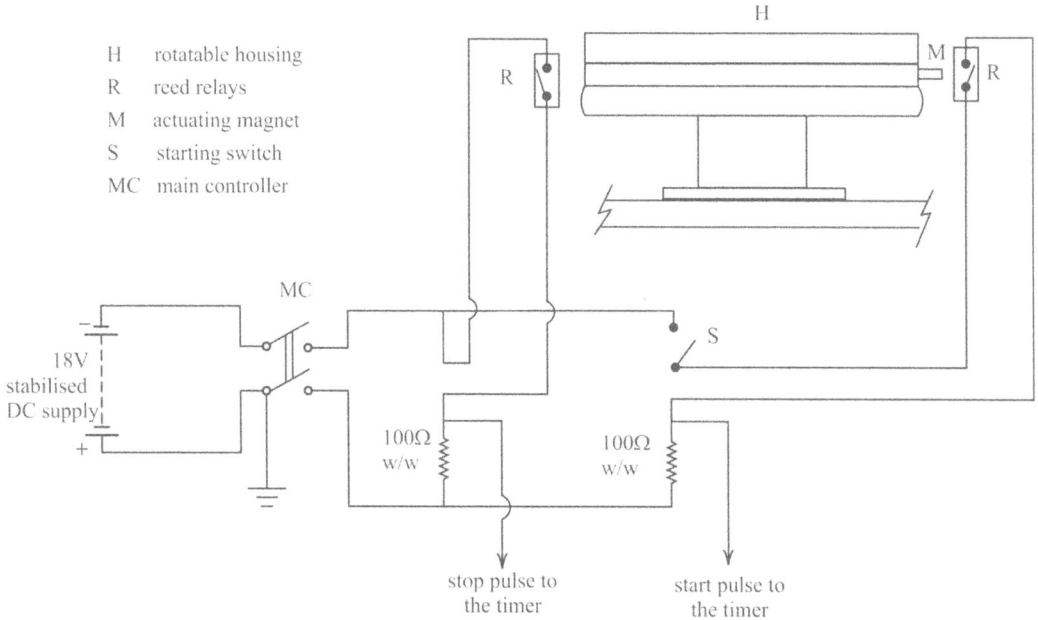

Fig.C1.8 : Schematic of speed measurement

Following the above scheme, the time interval for half revolution of the field housing could be measured to within ± 0.01 s and consistency of speed measurement within ± 2 %.

---

[9]The relative concentricity was ascertained by rotating the field system slowly (about ½ rpm) and at a low excitation measuring the developed torque that showed little variation over one complete revolution.

MEASUREMENT OF FLUX DENSITY

Search Coils for Flux Measurements

Search coils form a unique means for measurement of flux from which flux density can be derived, and these can be devised variously depending on requirements[10,11]. The choice of a set of search coils on the rotor of the machine to obtain a comprehensive flux density distribution was governed by various factors discussed later. However, a few novel constructional details are.

Airgap search coils

Considering the requirements discussed later in Chapter 2, the *radial* component of flux density in the *airgap* was required at two different radii. This could be accomplished by simply winding two full-pitch coils in the airgap adjacent to the external rotor surface, having their active lengths at two desired diameters. The alternative method would be to use one full-pitch coil giving the radial flux density at one of the radii and a single-turn coil with its plane on the same 'radius vector', giving an induced EMF proportional to the *difference* of the two radial flux densities. When added algebraically to the EMF of the first coil, this would yield the radial flux density value at the other radius.

The first method being more versatile with proper alignment of the search conductors and there being no 'risk' of stray 'pick-up', the induced EMFs proportional to both the radial components as well as to the difference depending on the terminal connections, was adopted. A sectional view of the location of search coils on either side of the annulus is given in Fig.C1.9.

With the constraint of a small airgap in the machine, being just 0.4 mm, very thin wire (49 SWG, 0.0375 mm OD) had to be used for winding the coils[12]. To achieve good mechanical alignment, a microscope having a magnification of 10 was used during the process of winding.

---

[10]See, for example,

 S.C.Bhargava: Electrical Measuring Instruments and Measurements (book), B S Publishers, 2013, Ch.12.

[11]Also,

S.C.Bhargava: The use of search coils for magnetic measurements, Int. J. Elect. Enging. Educ., Vol. 19, 1982, pp 45-52.

[12]This provided a much desired sharp contour for active coil sides.
Ibid.

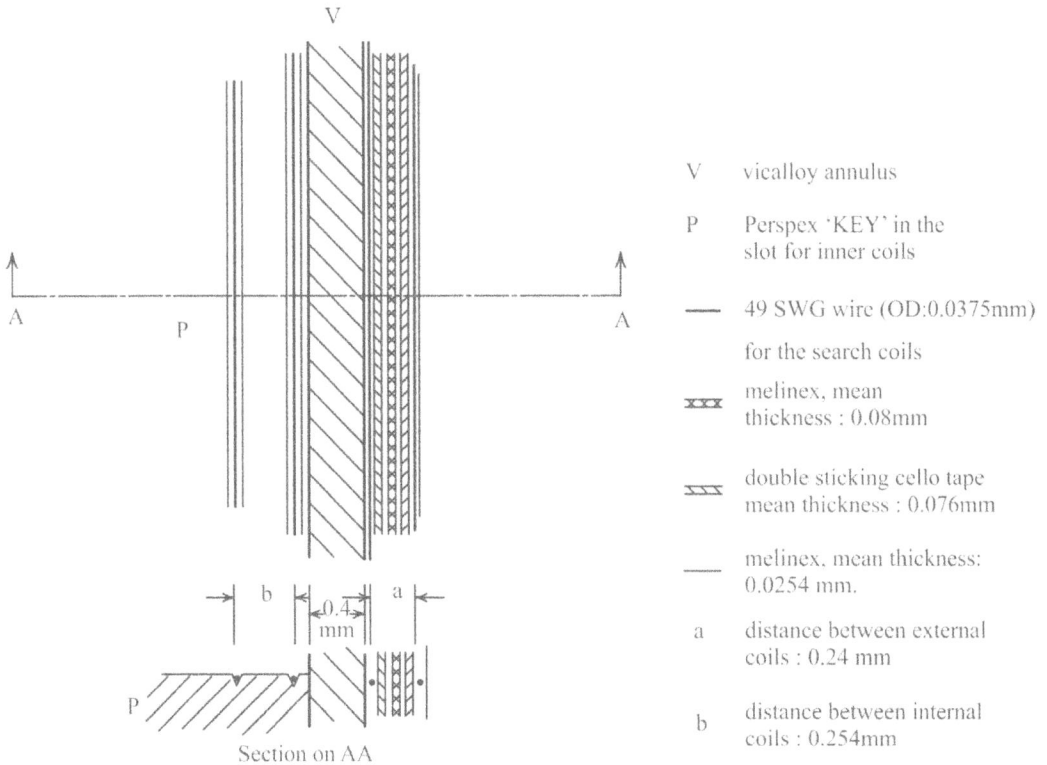

V   vicalloy annulus

P   Perspex 'KEY' in the
     slot for inner coils

——   49 SWG wire (OD:0.0375mm)
     for the search coils

xxx melinex, mean
     thickness : 0.08mm

xxx double sticking cello tape
     mean thickness : 0.076mm

——   melinex, mean thickness:
     0.0254 mm.

a   distance between external
     coils : 0.24 mm

b   distance between internal
     coils : 0.254mm

Section on AA

note : all dimensions converted from inches

Fig.C1.9 : Much enlarged sectional view of airgap search coils

Inner search coils

Two full-pitch search coils, similar to those in the airgap, were wound in the arbor region adjacent to the internal rotor surface and on the same radius vector as the airgap search coils to study the magnetic field distribution on the inner rotor surface; in this case the restriction on space requirement being not so critical.

The four full-pitch coils were thus on the same diametrical plane at four different known radii as depicted in Fig.C1.10.

Search Coils Round the Vicalloy Annulus

Two search coils were wound round the rotor annulus on the side matching the interpolar axis for miscellaneous tests. Specifically, one was a single-turn coil enclosing full annulus section, designed to measure induced EMF proportional to the 'entire' peripheral flux density inside the annulus. The other coil, of five turns, was wound threading two small holes (of 0.25 mm dia.), 12 mm apart along the axial length, or width, of the annulus. This coil was meant for ballistic measurements on vicalloy as detailed in Appendix I.

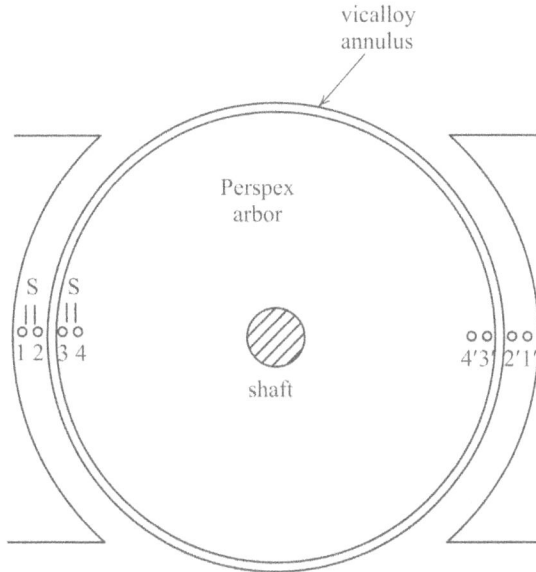

1-1', 2-2': coils in the airgap; 3-3', 4-4': coils in the arbor;
S: separation between coil(s) conductors (=0.076 mm)

Fig.C1.10 : Location of full-pitch search coils in the airgap and arbor

The details of all the search coils are summarised in Table C1.1.

Table C1.1: Various search coil details

Coil description	No. of turns	Conductor used	Size/Diametrical distance between active sides	Remarks
4 full pitch coils on the vicalloy rotor	One each	49 SWG, OD: 0.0015″ (0.0375mm)	(i) 3.2384″ (82.2554 mm) (ii) 3.2195″ (81.7754 mm) (iii) 3.174″ (80.6196 mm) (iv) 3.154″ (80.1116mm)	Mean diameter in order from the airgap towards the arbor. Active length : 25.4 mm
2 full pitch coils on the air-rotor	One each	Same	(i) 3.2384″ (82.2554mm) (ii) 3.2195″ (81.7754mm)	Active length : 25.4mm

Contd...

One coil round the vicalloy annulus for B_θ measurement	One	48 SWG, OD: 0.0019 " (0.0475mm)	25.4 mm × 0.4475 mm	Separation between the conductors : 0.4 mm (= thickness of the rotor annulus)
One coil enclosing the central portion of the annulus length	Five	49 SWG, OD: 0.0015 " (0.0375 mm)	12.0 mm × 0.4375 mm	For ballistic measurements. Separation between the 'turns': 0.4 mm, wound side-by-side.

Search Coils Instrumentation

Monitoring and measurement of EMFs induced in search coils posed un-common difficulties on account of extremely small output, being typically about 5 μV. The factors responsible for this were

- The main search coils were located in the *airgap* and the *non-magnetic* arbor region where the flux density was low to very low due to poor permeability of the material
- single-turn search coils were used to ensure accuracy of measurement of flux density at a 'point'
- the speed of rotation of the field system was only about 0.8 rps yielding small induced EMF
- the signals had to be recorded at a distance with the probability of attenuation and distortion en route

This necessitated special measures for instrumentation[13].

General Comments

At first sight the machine would appear to be large in mechanical shape and size. This is mainly a consequence of the space requirement for the two-pole excitation winding which was designed to carry huge number of turns so as to drive the rotor material well into magnetic saturation[14]. The overhang part of the winding are unavoidably large that would lead to appreciable leakage flux. See Fig.AIV.1, Appendix IV.

---

[13] *Ibid.*

[14] One of the many objectives of the research was to evaluate performance of a hysteresis machines under the conditions of saturation, even when the phenomenon would be non-linear.

The machine is also seen to have a short axial length compared to its diameter, the D/L ratio being about 3. This resulted from the overall design considerations and mechanical construction of the annulus. This aspect would reflect on a (two-dimensional) analysis vs. the experimental work. Nevertheless, machines with similar L/D rotor proportion are in use in practice[41], but in majority cases a nearly 1:1 ratio is observed in most machines.

2 : Tests on the Experimental Machine

2

Tests on the Experimental Machine

The common parameter for all the tests variously performed on the experimental machine was the applied mmf or excitation per pole, each test consisting of making observations or measurements with increasing excitation current in small steps, commensurate to the magnetisation of the rotor whilst preserving cyclic magnetic state in the machine. If the excitation was inadvertently set at a higher than the desired value, the machine was demagnetised and the test repeated.

TORQUE MEASUREMENTS

Variation of Torque with Excitation

A typical torque-excitation curve of the machine with the field system rotating at a *constant speed* of 0.8 rps is given in Fig.C2.1[1].

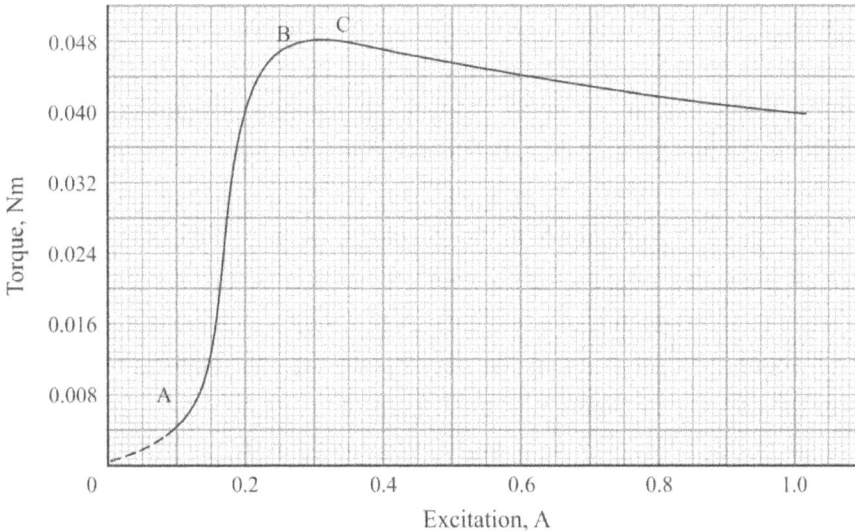

Fig.C2.1 : Torque-excitation curve of the hysteresis machine

---

[1]Observe the small magnitude of developed torque, disproportionate to the *mechanical* size of the machine. This is because the *volume* of the active material used, that is the vicalloy annulus, is too small. Just for the comparison, a hysteresis motor required to produce an output of one HP, may require an active material of the order of 0.015 m^3 operating at a flux density of 1 T and frequency of 50 Hz.

The shape of the curve can be somewhat related to the alternating hysteresis loss curve, or indirectly to the magnetisation curve, for vicalloy described in Appendix I. The results of measurement are reproduced in Table C2.1 for various measured flux density values in the rotor annulus, dealt with later in this chapter, and corresponding hysteresis loss computed from the areas of hysteresis loop at various excitations.

Table C2.1: Measured and computed torques for various excitations

Excitation	Time integrated mean B in the rotor on interpolar axis	Maximum flux density value, B_m	Alternating hysteresis loss corresponding to B_m	Torque equivalent to alternating hysteresis loss	Measured developed torque in the machine
A	tesla	tesla	$J/m^3/cycle \times 10^5$	Nm	Nm
0.1	0.055	0.07	0.075	0.0031	0.004
0.15					0.012
0.2	0.79	1.01	0.67	0.0275	0.04
0.25					0.048
0.3	1.02	1.3	1.137	0.047	0.048
0.4	1.09	1.39	1.137	0.047	0.048
0.6	1.12	1.43	1.137	0.047	0.044
1.0	1.14	1.45	1.137	0.047	0.04

The measured and so-calculated torque values for various excitations are plotted in Fig.C.2.2.

The torque at low excitations up to about 0.1 A (section OA in Fig.C2.1) increases slowly matching the initial part of the B-H curve, the loss due to hysteresis being small. At higher excitations (section AB in Fig.C2.1), the torque increases rapidly, almost linearly with current and corresponds to the steeply rising part of the loss curve as shown in Fig.C2.2. Finally, the torque tends to level off (section BC) in accordance with near-saturation stage in vicalloy; at higher excitations (> 0.4 A to 1.0 A) there is, in effect, a *reduction* of about 17% (at 1 A) in the developed torque[2]. This may be ascribed to non-uniform flux density distribution in the annulus due to saturation and slightly altered magnetic properties owing to heating at high excitation currents.

---

[2]Note that the calculated torque based on hysteresis loss corresponding to 'maximum' measured flux density would not show any reduction in torque as indicated by curve "b" in Fig.C2.2.

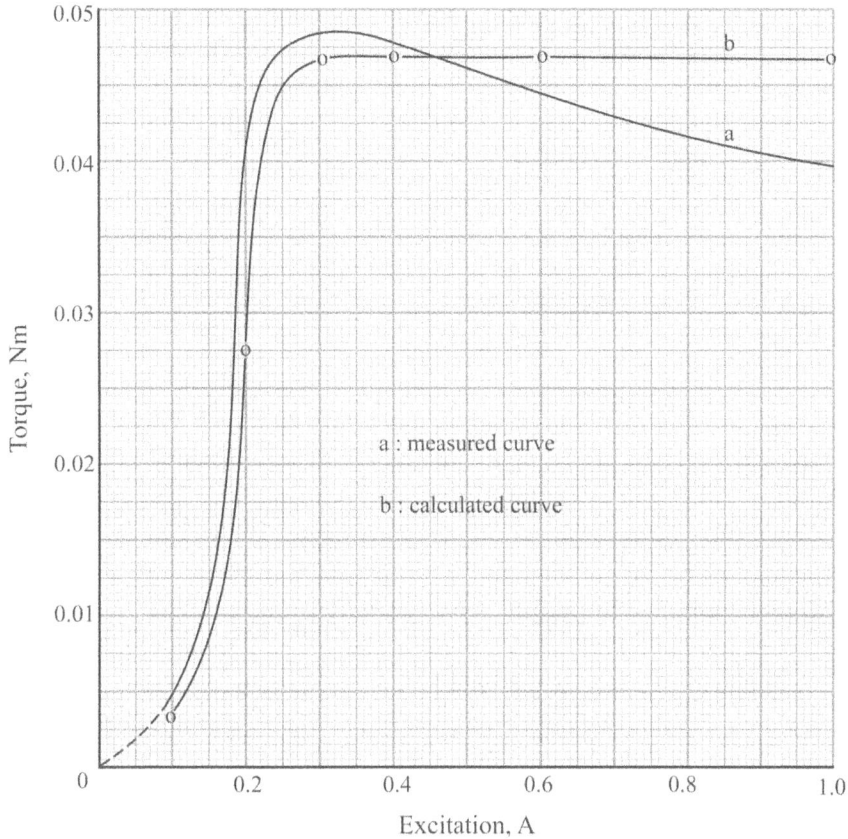

Fig.C2.2 : Measured and calculated torque-excitation curves

Variation of Torque with Speed of Rotation

The mechanically rotated magnetising field resulting in development of torque in the rotor would represent a non-synchronous operation of the machine. A test at varying speeds down to standstill at a given excitation current would demonstrate whether the machine operates in a manner similar to a 'normal' hysteresis motor at synchronous speed.

Accordingly, an excitation current of 0.3 A (corresponding to maximum developed torque) was used and the field system rotated at the highest speed of 2.5 rps. The speed was gradually reduced to standstill; this corresponding to the slip becoming zero in an actual motor. Whilst holding the field system stationary for a while, the torque was observed to be the same as during 'normal' operation. The torque dropped to zero when the field system was released. The entire operation is depicted in Fig.C2.3 by the vertical PQ.

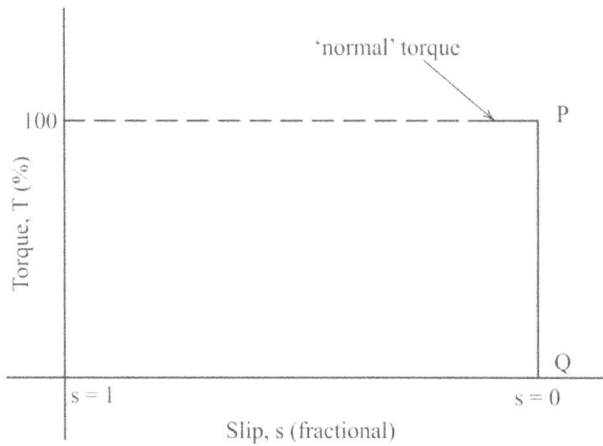

Fig.C2.3 : Synchronous operation of the hysteresis machine

The variation has the same shape as a part of the ideal torque-slip characteristic of a normal hysteresis motor, in the absence of both eddy currents and space mmf harmonics[3].

MEASUREMENT OF FLUX DENSITY

Measurement of flux density variously in the experimental machine formed an important requirement for the analyses at a later stage.

Requirements for Flux Density Measurement

With the assumption of little variation of magnetic field in the axial direction, the study of field distribution was confined to measurement in the radial and peripheral directions.

Radial flux density, B_r, distribution

The measurement of radial component of flux density was based on the "flux-cutting" rule applicable to the full-pitch coils described earlier. The EMF induced in the coil sides provided either space distribution of flux density at any instant or the variation with respect to time for a given point on the rotor; it was the former interpretation that would apply to analyses later. Also, an accurate diametrical positioning of the coil sides was essential as any "short-pitching" would eliminate certain harmonics from the induced EMF.

---

[3]Space mmf harmonics are present in the experimental machine; however, these are 'locked'' as discussed earlier having a common speed and do not cause a variation in the torque-speed characteristic.

Peripheral flux density, B_θ, distribution

The Usual Method of Measurement

The usual method for deriving peripheral flux density in the annulus would involve the use of a single- or multi-turn search coil enclosing the desired cross section. A similar coil placed in the air region or the airgap with one side in close contact with the annulus surface would result in an output proportional to the 'average' tangential component of the magnetising force, H_θ, across the plane of the coil. The EMF induced in such a coil would be proportional to the time-rate-of-change of the flux linking the coil at the given speed, and therefore time integration of the coil output would provide the variation of H_θ.

Since area of the coil has to be finite to yield measurable output, the accuracy of the method would depend on *width* of the coil in a plane perpendicular to the external surface of the annulus. Also, to define a sharp contour, the coil would be of single turn, wound using very thin wire[4].

Validity of the method

The validity of the assumption that a narrow (airgap) coil yields B_θ (or H_θ) on the longitudinal axis of the coil after time integration can be established by considering the induction of EMFs on the basis of the flux-cutting rule applied to radial components of flux density as configured in Fig.C2.4.

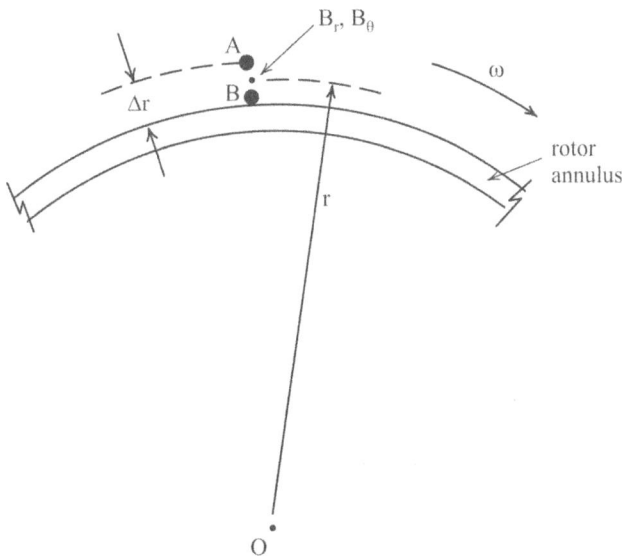

Fig.C2.4 : Configuration of search coil conductors in the airgap

---

[4]*Ibid.*

Let the radial and peripheral components of flux density at the *mean* radius, r, of the coil, having coil sides at A and B, Δr apart, be B_r and B_θ. If the difference in the values of B_r at A and B is ΔB_r, then the EMFs induced in the coil sides will be

$$e_A = \omega l \left(r + \frac{\Delta r}{2} \right) \left(B_r + \frac{\Delta B_r}{2} \right) \tag{C2.1a}$$

$$e_B = \omega l \left(r - \frac{\Delta r}{2} \right) \left(B_r - \frac{\Delta B_r}{2} \right) \tag{C2.1b}$$

where l is the axial length and ω the angular speed of rotation of the coil(sides) relative to the field.

The output from the (time) integrator will be

$$(\text{output})_{int} = \int (e_A - e_B) \, dt$$

and the (apparent) peripheral flux density, B_θ', related to the output given by

$$B_\theta' = -\frac{1}{a} \int (e_A - e_B) dt, \text{ a being area of the coil } (= l\Delta r)$$

Substitution for e_A and e_B, followed by some simplification gives

$$B_\theta' = -\int \left(r \frac{\Delta B_r}{\Delta r} + B_r \right) \omega \, dt \tag{C2.2}$$

or

$$B_\theta' = -\int \left(r \frac{\Delta B_r}{\Delta r} + B_r \right) d\theta \tag{C2.3}$$

Now at the mean radius

$$\overline{\nabla} \cdot \overline{B} = \frac{1}{r} \frac{\partial B_\theta}{\partial \theta} + \frac{\partial B_r}{\partial r} + \frac{B_r}{r} = 0$$

or

$$-\left(r \frac{\partial B_r}{\partial r} + B_r \right) = \frac{\partial B_\theta}{\partial \theta} \tag{C2.4}$$

If therefore the width of the coil is 'sufficiently' small, the partial derivatives of eqn.(C2.4) can be replaced by the incremental values ΔB_r, ΔB_θ, Δr and

$$B_\theta' = B_\theta \tag{C2.5}$$

Errors Involved and Further Limitations

A drawback with the calculated value of B_θ from the integrated search coil output is that it is essentially a *mean* value that might cause any analysis involving B_θ on the rotor *surface* erroneous.

In addition, it is seen that the coefficients of sine and cosine terms of the high-order harmonics in the time-integrated B_θ waveforms may differ appreciably from the correct values.

To explain this, consider a magnetic scalar potential distribution in the air region

$$\phi = A_n \, r^n \sin n\theta \tag{C2.6}$$

where n is the order of harmonic and A_n the coefficient of the nth term.

The correct expressions for the B_r and B_θ components, derived from eqn.(C2.6), are

$$B_r = \mu_0 \, H_r = -\mu_0 \frac{\partial \phi}{\partial r} = -A_n \, \mu_0 \, n \, r^{n-1} \sin n\theta \tag{C2.7a}$$

and
$$B_\theta = \mu_0 \, H_\theta = -\mu_0 \frac{1}{r}\frac{\partial \phi}{\partial \theta} = -A_n \, \mu_0 \, n \, r^{n-1} \cos n\theta \tag{C2.7b}$$

using $\overline{H} = -\overline{\nabla}\phi$.

If the coil sides A and B of the search coil (Fig.C2.3) are located at radii a and b, respectively, then B_θ at the mean radius, $r = (a + b)/2$ will be

$$B_\theta = -A_n \, \mu_0 \, n \left[\left(\frac{a+b}{2}\right)^{n-1} \right] \cos n\theta \tag{C2.8}$$

and the radial flux densities at a and b given by

$$B_{r_a} = -A_n \, \mu_0 \, n \, a^{n-1} \sin n\theta \tag{C2.9a}$$

$$B_{r_b} = -A_n \, \mu_0 \, n \, b^{n-1} \sin n\theta \tag{C2.9b}$$

For B_θ obtained from time integration of the *difference* of the induced EMFs proportional to B_{r_a} and B_{r_b}, it can be shown using eqn.(C2.3) that the *apparent* peripheral flux density at $r = (a + b)/2$ is given by

$$B'_\theta = -A_n \, \mu_0 \, n \left[\left(\frac{1}{n}\frac{a^n - b^n}{a - b}\right) \right] \cos n\theta \tag{C2.10}$$

A comparison of eqns.(C2.8) and (C2.10) would show that the coefficients of $\cos n\theta$ terms are different for the two cases and depend on the order of harmonic, n. To realise the extent of error involved, the expressions within the brackets were evaluated for various values of n and are reproduced in Table C2.2. The error becomes progressively worse with the increase of harmonic order. Also, error for the same order of harmonic would be worse still for larger values of $|a-b|$[5].

Table C2.2: Apparent peripheral flux density

a = 1.0, b = 0.99

Terms in [ ] bracket in eqns. C2.8 and C2.10	Order of harmonic, n→	1	3	5	7	9	19	49	99	199	499	999
$\left(\dfrac{a+b}{2}\right)^{n-1}$		1.0	0.9900	0.9801	0.9704	0.9607	0.9138	0.7861	0.6119	0.3707	0.0824	0.0067
$\dfrac{a^n - b^n}{n(a-b)}$		1.0	0.9900	0.9802	0.9705	0.9609	0.9149	0.7936	0.6366	0.4345	0.1991	0.1001

Field Theory Integration

An alternative, rigorous approach was adopted to eliminate the foregoing limitations; the approach being based on field theory.

If the *radial* flux density waveforms (or variation) were determined accurately at two different radii in the region and expanded into Fourier series, then the B_θ waveforms could be derived using the coefficients of the harmonic terms in these series[6]. Since the radial flux density waveforms on the external and internal rotor surfaces would be obtained using two full-pitch search coils as described earlier, similar waveforms at another point on the same radius vector could be derived using two other coils suitably located. To achieve better accuracy, the two coils on each side of the rotor annulus were connected electrically to provide the EMFs from the inner coil and an EMF proportional to the difference of radial flux densities, ΔB_r. The two EMFs were added to give the EMF in the outer coil. A schematic of the arrangement of various coils is given in Fig.C2.5.

---

[5]It would be seen later that the harmonics contribute appreciably to the torque developed in the experimental machine; hence a detailed account of the effect of harmonics in the values of flux density components.

[6]See Appendix III for details of derivation and computerisation.

P   Perspex arbor
V,V   vicalloy rotor section
S   screened rotary switch
A1   diff. input op amp, gain unity
A2   diff. input ins. amp, gain 100
A3   diff. input ins. amp, gain 100

G1
G2 } UVR galvanometers
G3

CC   two sets of coils

CP   common points

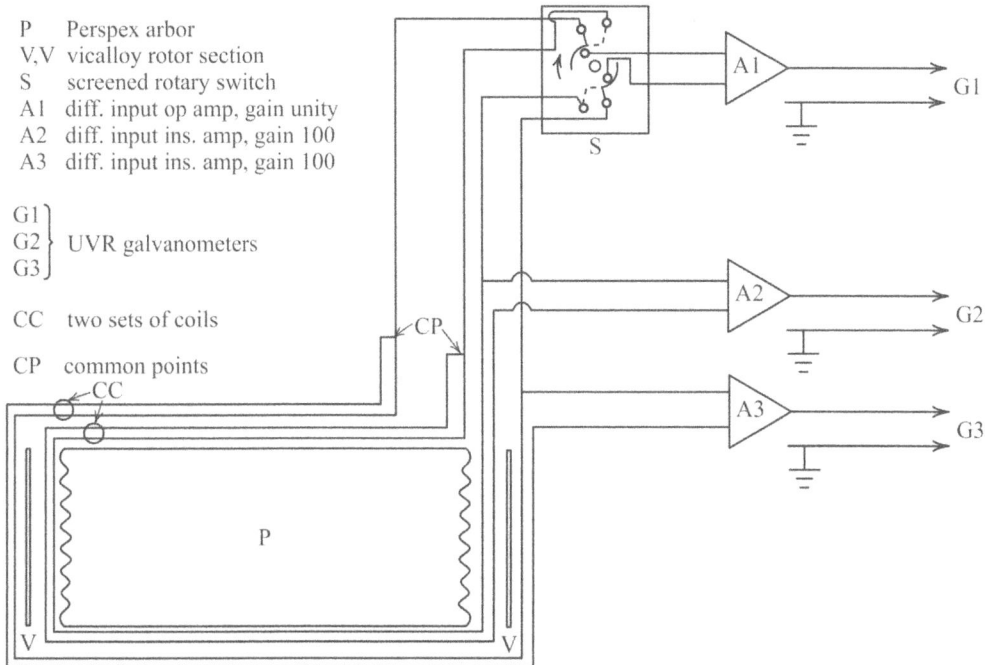

Fig.C2.5 : Schematic of full-pitch search coils

Radial flux density measurements

B_r and ΔB_r Measurements with "Air Rotor"

Some preliminary tests were performed on the machine with the field system excited, *but held stationary* and the assembly behaving like a non-linear magnetic circuit[7]. The initial stage consisted of experiments with a non-magnetic, non-metallic rotor, identified as "air rotor", that consisted of tufnol, machined to dimensions identical to the rotor with vicalloy annulus[8]. The rotor was provided with full-pitch search coils identical to those on the vicalloy rotor for measurement of B_r and ΔB_r flux densities. Such a rotor, having unit relative permeability, would result in a flux distribution similar to that due to low-permeability vicalloy rotor, *but without the influence of spatial hysteresis*. The recorded waveforms of B_r and ΔB_r at 0.1 and 0.3 A[9] excitations are shown in Fig.C2.6[10], characterised by huge peaks in flux density values at the pole tips. These peaks which

---

[7]The non-linearity arising on account of saturation of pole tips, owing to design of the field system, as reflected in the measured waveforms.

[8]See Appendices IV and V.

[9]These excitation currents are representative of non-hysteretic and pronounced hysteretic conditions in the vicalloy rotor and were chosen to typify the various measured waveforms all through and to make comparisons possible.

[10]Showing only relative trend; to varying scale.

also occur with the vicalloy rotor, both in B_r and B_θ waveforms reported next, are caused by sudden change in the pole-tips geometry and are significant to the machine operation as a whole. The low permeability of the rotor, either "air" or vicalloy, results in low radial flux density under the pole (centre); the low flux density is not due solely to low relative permeability, however, but is also ascribed to small radial thickness of the rotor annulus and may be observed even with high-permeability material if the annulus were very thin.

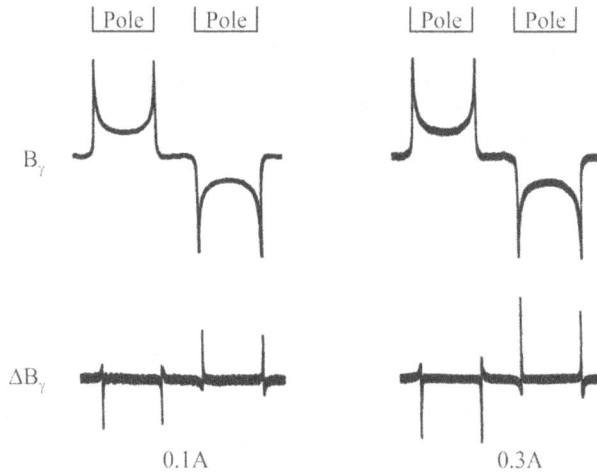

Fig.C2.6 : B_r and ΔB_r flux density waveforms with the "air" rotor

In the absence of magnetic hysteresis, the waveforms were analysed using Teledeltos paper plots and conformal transformation[11], and revealed that some of the characteristic features of the machine, observed during its operation, were due to its design and geometry. Although the analogue models were intended as a guide, the actual measurements of flux density showed good correlation.

B_r and ΔB_r Measurements with Vicalloy Rotor

B_r and ΔB_r in airgap

The oscillograms of these waveforms are shown in Fig.C2.7[12]. A comparison of these with the waveforms for the air rotor bring out the role played by spatial hysteresis of the vicalloy rotor. When viewed in the field-system reference frame, the flux in either pole is concentrated in the lagging pole tip whilst relatively weak at the leading pole tip as illustrated in Fig.C2.8, rotor annulus thickness being exaggerated for clarity.

---

[11]See Appendix IV.

[12]It is observed that the 'base' of ΔB_r waveforms is not sharp, obscured by the amplifier noise and slight 50 Hz pick-up from the mains, the signal being too small owing to B_r values at the two radii being almost equal at any 'point' in space.

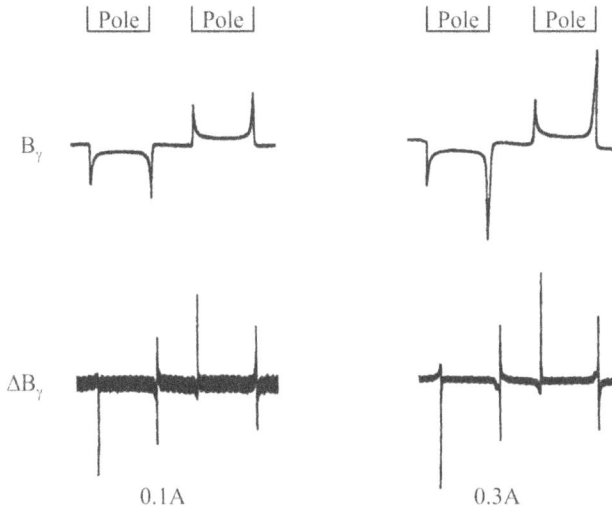

Fig.C2.7 : B_r and ΔB_r waveforms in the airgap of vicalloy rotor

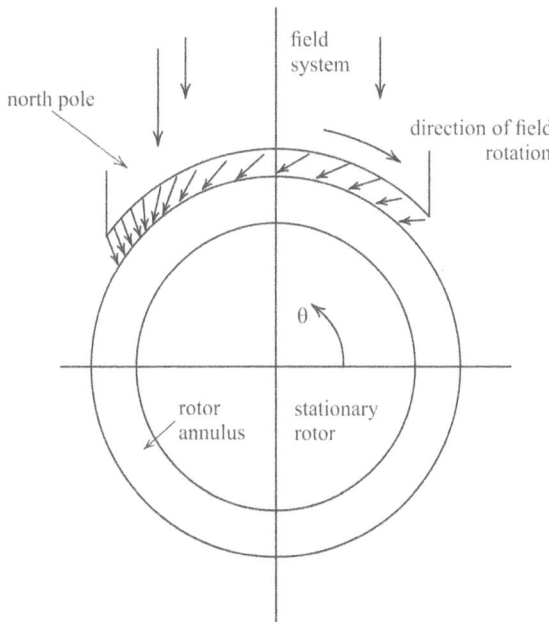

Fig.C2.8 : Uneven flux distribution in the poles of the field system

The uneven flux distribution in the pole tips leads to unequal heights of peaks in the measured waveforms. Note that the sharp peaks in the various waveforms do not necessarily occur opposite the pole tips since the B_r waveforms though very much similar in shape, differ appreciably at certain positions in space. This makes it essential to record B_r AND ΔB_r variation instead of just two B_r waveforms.

Time-integrated B_θ Waveforms in the Airgap

These waveforms were recorded to indicate general shape of the peripheral flux density variation influenced by rotor hysteresis and are shown in Fig.C2.9; the waveforms are markedly affected by drift in the circuit, accentuated by gain of amplification[13].

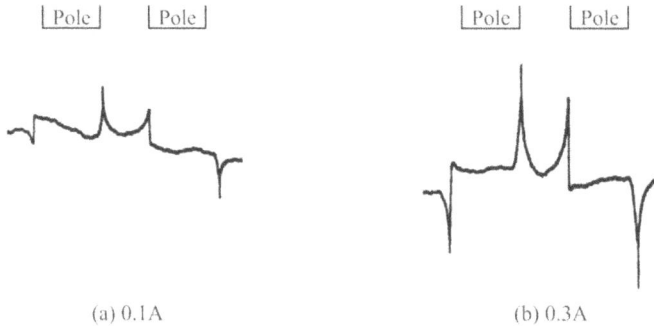

(a) 0.1A   (b) 0.3A

Fig.C2.9 : Time-integrated B_θ waveforms in airgap of vicalloy rotor

B_r and ΔB_r Measurements in the Arbor Region

These waveforms are shown in Fig.C2.10. The B_r waveforms indicate that the spatial hysteretic effect, reflected in the difference of heights of peaks at the pole tips, is not present to the same degree as in the airgap. The waveforms thus approach closely the B_r distribution of the air rotor having near symmetrical peaks. Nevertheless, an appreciable flux crosses into the arbor region for the amplitudes of the waveforms are comparable to those in the airgap. However, the peaks though still occurring near the pole tips are attenuated both in B_r as well as ΔB_r waveforms which means that the rotor shields the arbor region to an extent from higher-order harmonics despite its poor relative permeability.

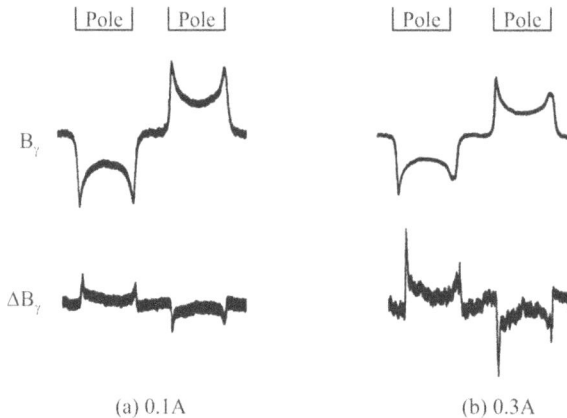

(a) 0.1A   (b) 0.3A

Fig.C2.10 : B_r and ΔB_r waveforms in the arbor region with vicalloy rotor

---

[13] A pre-amplifier with a gain of 100, connected to an integrator comprising a Burr-Brown operational amplifier with very large CMMR: overall gain being 10^4.

Peripheral Flux Density Waveforms for Vicalloy Annulus

These waveforms were obtained from the single-turn search coil wound round the vicalloy annulus and its time integration, shown in Fig.C2.11; the drift being comparatively small and harmonic content being low, the error due to time integration was negligible and recordings more 'accurate'.

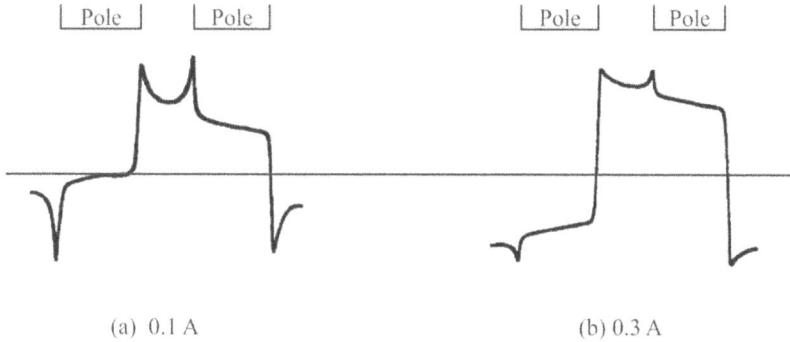

(a) 0.1 A   (b) 0.3 A

Fig.C2.11 : Time-integrated (or average) peripheral flux density, B_θ, waveforms in vicalloy annulus

In the waveforms the peaks occur almost directly opposite the pole tips at low values of excitation because the flux distribution within the rotor is governed by low permeability and negligible hysteresis. At an excitation of 0.3 A, spatial hysteresis is more pronounced and the flux distribution is nearly constant as indicated by the "square-shape" of the waveform. These waveforms were later used to derive a torque-excitation curve based on pure alternating hysteresis loss.

Concluding Remarks

The tests described above elucidate a number of characteristic features of the experimental machine. In the range of 0.15 A to 0.25 A, the developed torque is proportional to excitation, making the machine useful in control systems. The torque has a maximum value very nearly equal to the theoretical maximum available on the basis of alternating hysteresis loss in the vicalloy annulus. The loss of torque at higher excitations (>0.3 A) can be attributed to alteration of magnetisation cycle in the rotor as also general modification of flux distribution in the machine including intense saturation[14]. There would also be some effect on the developed torque due to "rotational hysteresis" as discussed later.

---

[14]To eliminate errors arising from any excitation current variation and change in speed, the B_r and ΔB_r waveforms for the external and internal rotor surfaces were recorded *simultaneously* on a UVR.

The waveforms formed the basis for development of the **qualitative theory of operation** and arrive at quantitative analysis of power flow in the machine as described in Chapter 4.

3 : Computed Magnetic Flux Density Variation

3

Computed Magnetic Flux Density Variation

The non-linear B/H relationship for vicalloy and even steel when saturated, and complex machine geometry would make it difficult to derive magnetic flux density variation from first principles. Instead, the measured *radial* flux density waveforms are used in conjunction with Maxwell's equations to derive the various waveforms in the air spaces; in particular the peripheral flux density waveforms are obtained both for the vicalloy and air rotor. A characteristic feature of these waveforms is the existence of a unidirectional flux density under the poles caused by spatial hysteretic property of vicalloy annulus, indicating a shift in the 'pole axis' during machine operation in the direction of rotation of the field system.

Computation of Waveforms

General considerations

The details of proceeding from the recorded B_r and ΔB_r waveforms to derive expressions for the peripheral flux density variations as well as magnetic scalar potential are given in Appendix V. Since the (measured) waveforms contain huge peaks towards the pole tips, a unique procedure was devised for harmonic analysis[1].

Order of harmonics

The first step in computation was to decide on the highest order of harmonic where the Fourier series for B_r and ΔB_r waveforms could be truncated and yet achieve desired accuracy of harmonic analysis considering the peaky waveforms encountered in measurements. After repeated computations, it was noted that the changes in the results – peak values of either radial or peripheral flux density and power flow – was less than

---

[1]See, for example,

S.C.Bhargava: A new method for Fourier analysis of peaky waveforms, Int. J. Math. Educ. Sci. Technol., Vol. 17, No.6, 1986, pp 683-692.

and

S.C.Bhargava: A unique method for Fourier analysis of distorted periodic waveforms encountered in engineering problems, BHEL Journal, Vol.7, No.1, 1986, pp 35-41.

See also Appendix VI.

0.1% for harmonic order over 199, that is 100 harmonic terms since only odd harmonics would be present.

All subsequent computations were therefore carried out for 100 harmonics[2].

Computed Radial Flux Density Waveforms

Although waveforms of radial flux density, B_r, were available from the outputs of the full-pitch search coils, both in the airgap and arbor region, 'new' waveforms were evaluated for two main reasons. Firstly, the new waveforms were calculated at the same values of angular position as all other waveforms which made comparison much easier. Secondly, it was essential to have the variation known on the internal and external rotor *surfaces* rather than at any other radii.

B_r on the external rotor surface

Considering general shape of radial flux density with vicalloy rotor having peaks of unequal heights at pole tips, with the field system rotating in a clockwise direction and measurements made on search coils placed on the stationary rotor, the peak with smaller height would occur when pole tip A moves past the coil as indicated in Fig.C3.1(same as Fig.C2.8). In what follows, this pole tip would be referred to as "leading" pole tip with the angle, θ, measured anticlockwise in the rotor reference frame; the other pole tip, B, is identified as "lagging" pole tip.

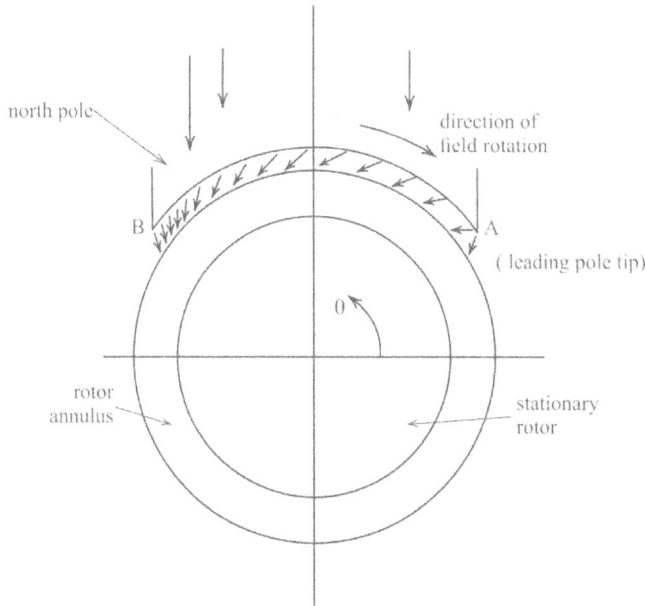

Fig.C3.1 : Rotating field system and angular position of the rotor

---

[2]As a check on computation and input data, the peak value of radial flux density was computed and found to be very nearly equal to the peak value of the measured waveform.

The computed B_r waveforms on the external rotor surface for excitation currents in the range 0.1 A to 1.0 A are shown in Fig.C3.2. Whilst generally representative of hysteretic conditions in the rotor, the relative heights of the peaks at the pole tips undergo distinct changes at increasing excitations; also, the waveforms under the poles are almost 'flat' up to an excitation of 0.3 A, but are seen to be smoothly curved at higher currents.

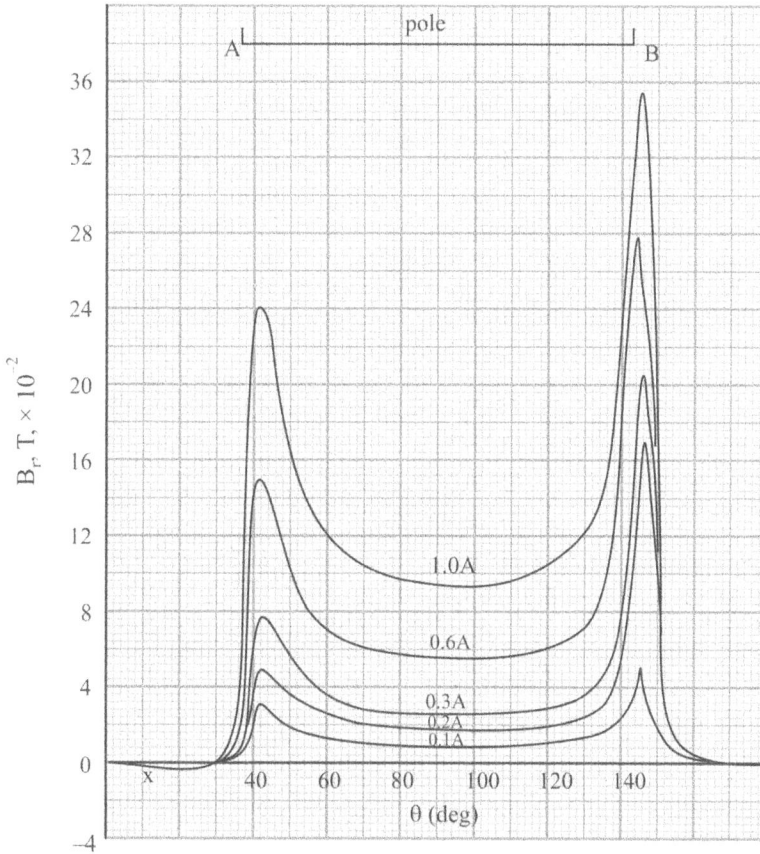

Fig.C3.2 : B_r variation on the external rotor surface at increasing excitations

This means that a good proportion of flux reaches the rotor from the central part of the pole at higher excitations (>0.3 A) than from the pole tips; the reverse being true at low excitations. Considering the flux entering the rotor from the pole, there is a noticeable tendency for some of this flux to leave the external rotor surface in the interpolar region at excitations of 0.6 A and 1.0 A as shown by a slight reversal in sign of the flux density at the interpolar axis (X in Fig.C3.2). This is due mainly to the intense saturation of pole tips and resulting low permeability of the surface in this region.

A trend clearly visible in the waveforms is the rapid growth of the peak at pole tip B relative to that at A in the current range 0.1 A – 0.3 A, the increase becoming comparable at higher currents; the degree of unbalance in the B_r waveform being almost proportional to the developed torque as discussed later. To some extent, this is due to the overall flux distribution from the pole, being less affected by the spatial hysteresis of the rotor.

B_r on the internal rotor surface

The waveforms of radial flux density variation on the internal surface of the vicalloy and air rotor are depicted in Fig.C3.3 for excitations of 0.1 and 0.3 A, being similar in shape and comparable.

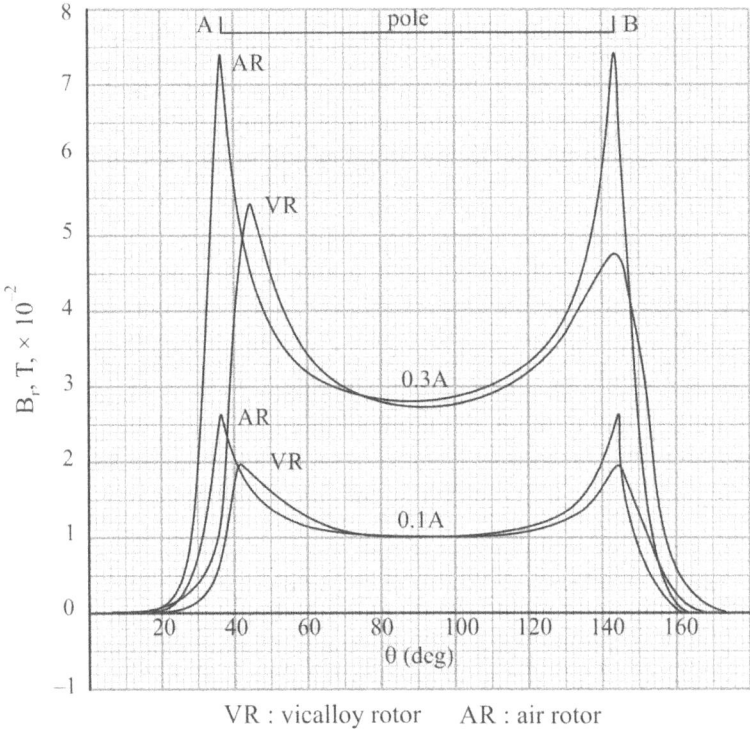

VR : vicalloy rotor   AR : air rotor

Fig.C3.3 : Radial flux density variation on internal surface

At 0.1 A excitation both waveforms appear symmetrical in terms of location and height of peaks, indicating nearly negligible spatial hysteresis. In contrast there is a marked difference at 0.3 a excitation: the variation for the air rotor is still symmetrical whereas that for the vicalloy rotor has peaks of different heights. This 'apparent' spatial hysteresis effect with the vicalloy rotor is due to rotor magnetisation and results in a power flow through the inner surface as brought out later[3] Note, also, that the larger peak now occurs opposite the leading pole tip, A, instead of opposite the lagging tip in the airgap.

---

[3]Described in Chapter 5.

Computed Peripheral Flux Density Waveforms

External rotor surface

The B_θ waveforms on external surface for both air and vicalloy rotors are illustrated in Figs.C3.4 and C3.5 together with the computed B_r waveforms.

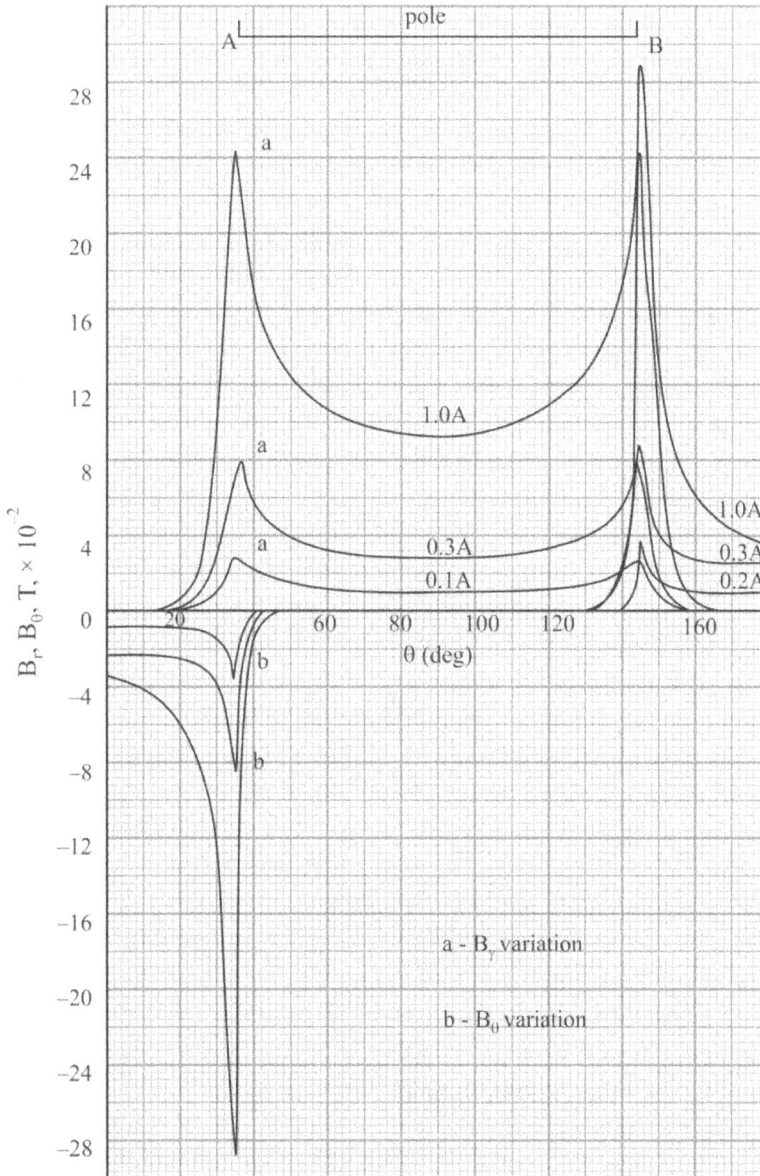

Fig.C3.4 : B_θ variation with air rotor

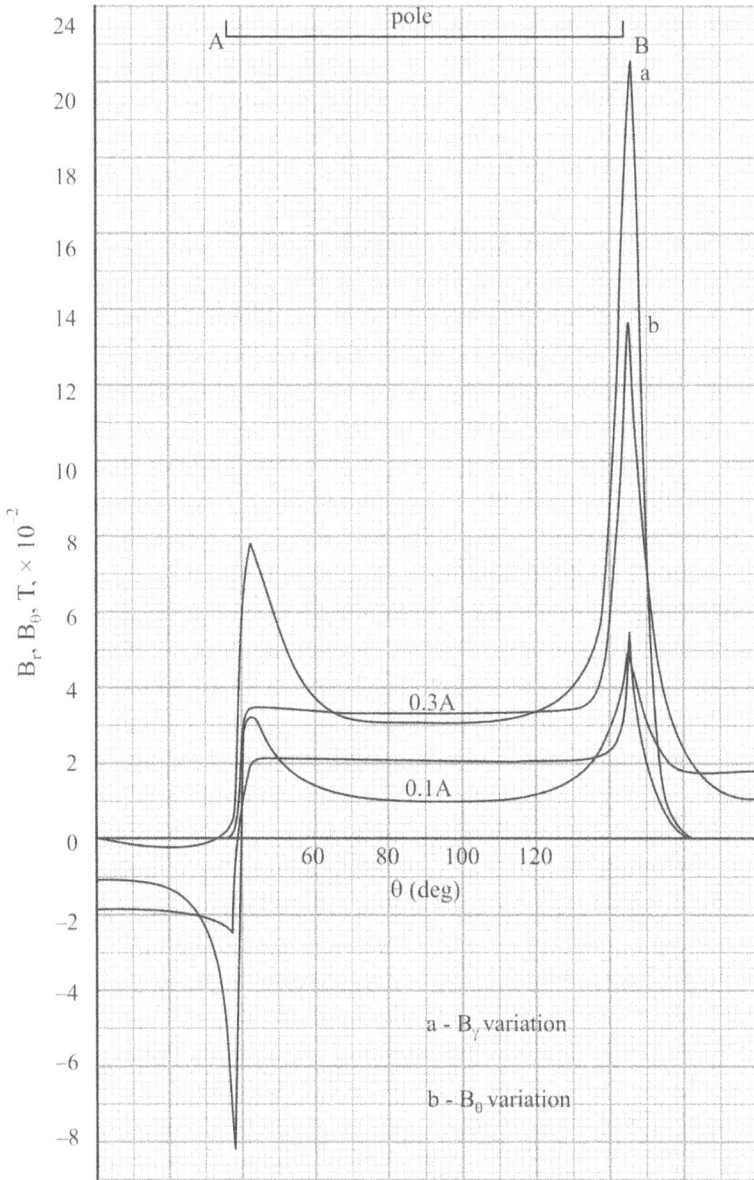

Fig.C3.5 : B_θ variation with vicalloy rotor

With the air rotor the B_θ and B_r waveforms are symmetrical and displaced in space by $90°$ (or $\pi/2$). This depicts the non-hysteretic condition and corresponds to zero power flow over a closed cylindrical surface. The shape of waveforms follows nearly the same trend from 0.1 to 1.0 A excitations.

In the case of vicalloy rotor (Fig.C3.5), an important feature of the B_θ waveforms at *all* excitations is the non-zero value of flux density under the poles which is practically constant over the pole arc. The existence of a non-zero, unidirectional B_θ under the pole is a sufficient but not necessary mathematical condition for net power flow and hence production of torque. This aspect is dealt with later in Chapter 5.

Another significant feature is that reversals in the waveforms occur opposite the *leading* pole tips. With reference to the flux leaving the pole, this means that the pole axis has shifted in space in the direction of field-system rotation; the flux now bifurcating near the leading pole tips instead of at the centre[4]. The flux at the pole tip can flow either to the opposite pole or to the yoke, but with the poor permeability of the rotor, further affected by intense magnetic saturation, the flux tends to leak away into the interpolar region towards the yoke. The approximate conditions pertaining to other common electric machines vis-à-vis the experimental machine are shown in Fig.C3.6; the inclined flux lines in the airgap of the latter having been deduced from the waveforms of Fig.C3.5 and discussed further later[5].

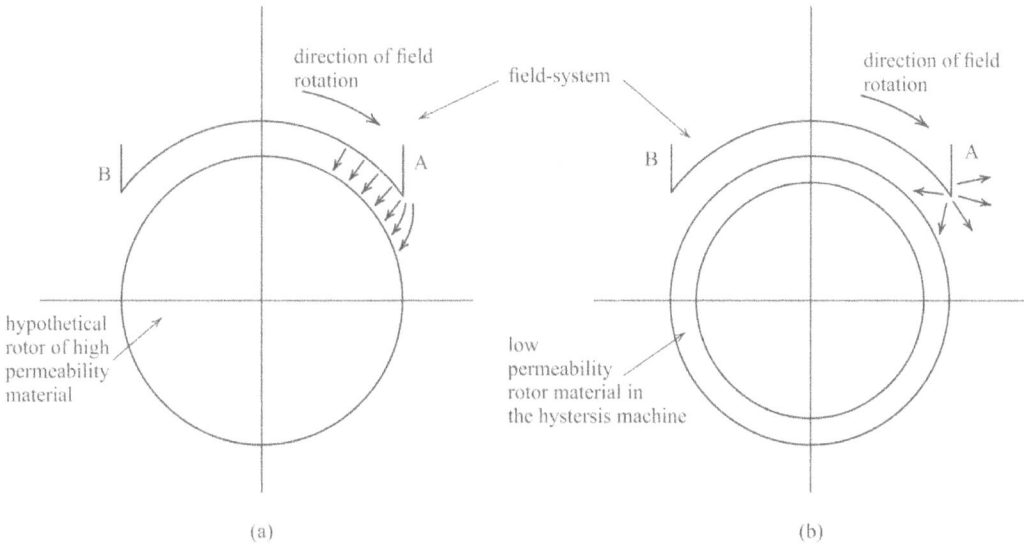

Fig.C3.6 : Airgap flux in normal and experimental machine

Internal rotor surface

The waveforms of the internal surface, given in Fig.C3.7, differ in two main respects from those on the external surface. Firstly, the values of B_θ on the interpolar axis are

---

[4]This effect is a special feature of the experimental machine since the location of flux division even at very low excitation is dictated by the sharp-edge pole tips.

[5]For a rigorous magnetic field calculation on account of saliency of the field system, see M.J.Jevons: Magnetic field calculation for a salient-pole hysteresis coupling, COMPUMAG, Oxford, England, 31 March-2 April, 1976.

much higher than the corresponding values on the external surface. Although this may appear improbable at first sight, a qualitative explanation can be extended as follows:

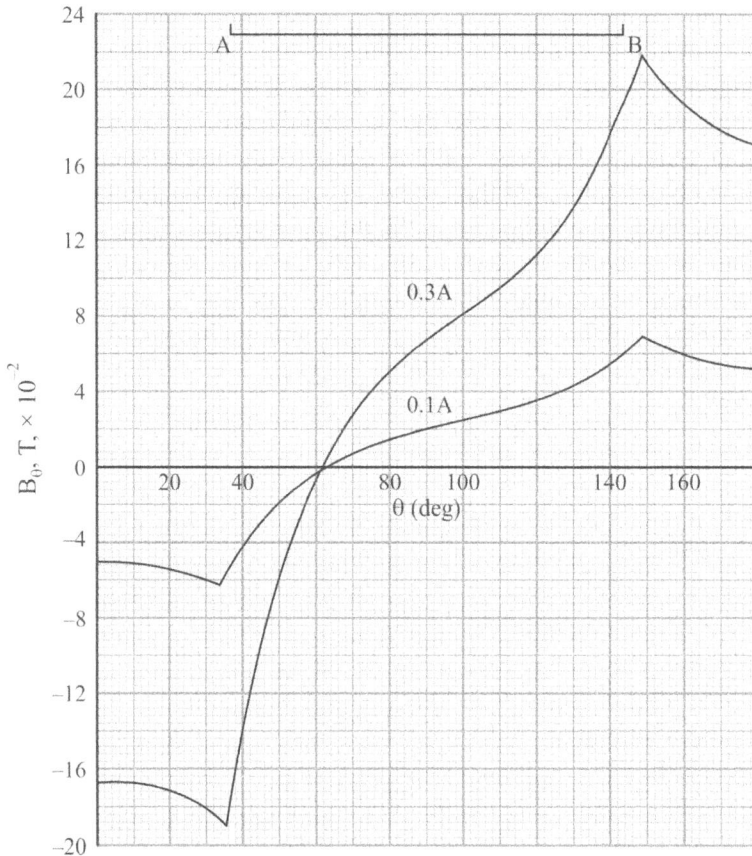

Fig.C3.7 : Peripheral flux density variation on internal rotor surface

Consider the field system of the machine without the vicalloy rotor (or with the air rotor). At low excitations, the flux distribution region between the pole surfaces takes the form shown in Fig.C3.8, also confirmed by iron-filing patterns later[6]. With the concentration of flux at the pole tips, the flux between the tips will clearly take the path such as a b c or a b' c, the path(s) of low reluctance, rather than a b" c which corresponds to the external surface of the air rotor. This would result in a higher flux density in the inner region, represented by Ob (or Ob') than Ob".

---

[6]See Appendix III.

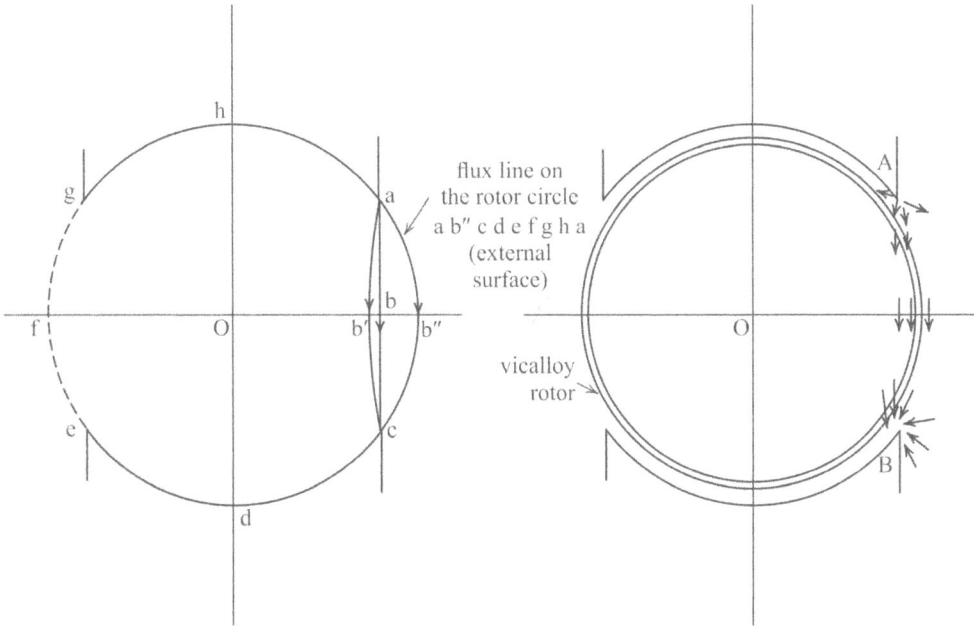

Fig.C3.8 : Flux pattern between poles without rotor   Fig.C3.9 : Flux pattern with vicalloy rotor

A similar situation occurs with the vicalloy rotor, Fig.C3.9, with great concentration of flux at pole tips. In particular, the flux density at the lagging pole tip B is always much higher than at A during the machine operation, the direction of flux lines at key positions being as indicated by small arrows. The presence of a thin rotor forces most of the flux leaving A in the direction of B to pass through the vicalloy and the inner region rather than take the path along the external rotor surface. A good proportion of the flux on the external pole surface in the interpolar region leaks away towards the yoke, especially from the lagging pole tip B; again confirmed by iron filing patterns[7].

The other difference between the waveforms on the two rotor surfaces is the point at which the peripheral flux density reverses. On the internal surface this lies between the leading pole tip and the pole axis; on the external surface this occurs between interpolar axis and pole tip A.

Resultant Flux Density Distribution

The resultant flux density variation was obtained by vectorially combining the B_r and B_θ values at any point in space, aimed to study the direction of flux lines entering different points on the rotor surface. These are best displayed as polar plots to provide a clear picture of flux distribution in the regions of interest, both in magnitude and direction, and are shown in Figs.C3.10 and C3.11 for representative currents of 0.1 A and 0.3 A.

---

[7]See Appendix IV.

Plots for the air rotor

The variation for the air rotor is depicted in Fig.C3.10. It is seen that the flux in the airgap under the pole surface is radial but changes direction at the pole tips to take a shorter path to the opposite pole, the resultant flux density being a maximum and having same value opposite the two pole tips. There is little alteration in the trend of variation at the two excitations. The plots match the flux distribution arrived at analytically and by iron-filing analogues[8]

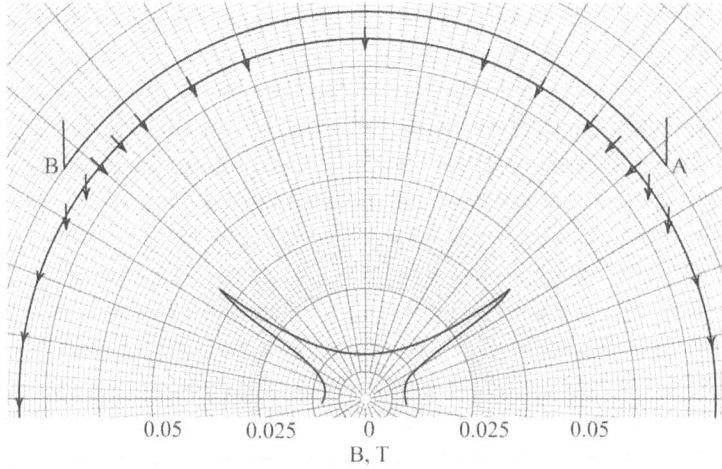

(a) 0.1 A excitation

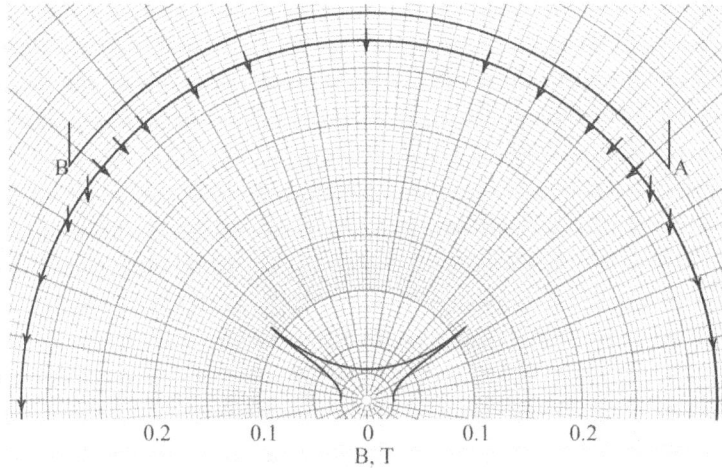

(b) 0.3 A excitation

Fig.C3.10 : Resultant flux density distribution: air rotor

---

[8]See Appendices III and IV.

Plots with the vicalloy rotor

These plots, shown in Fig.C3.11, are very different from those for the air rotor. The flux lines are no longer radial in the airgap, but inclined in a peculiar manner owing to non-zero peripheral flux density under the pole surface, a characteristic of this machine.

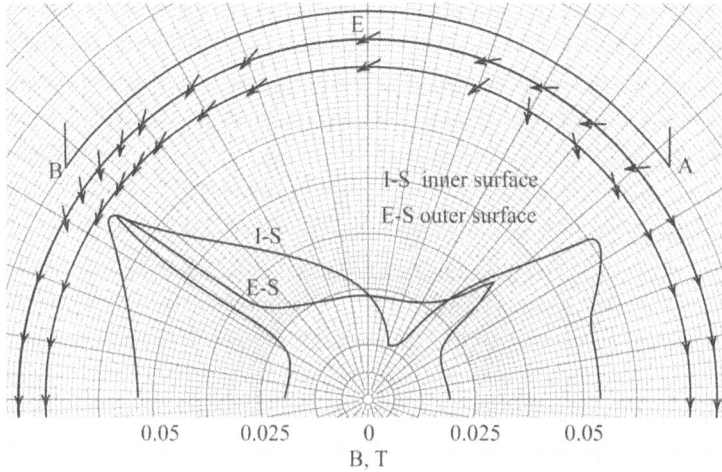

(a) 0.1 A excitation

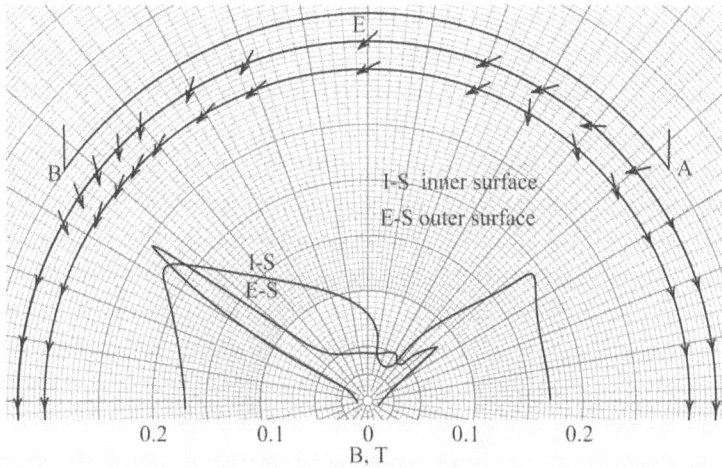

(b) 0.3 A excitation

Fig.C3.11 : Resultant flux density distribution: vicalloy rotor

The angle of inclination of flux at different points on the external rotor surface under the pole is in the range of $20°$ to $50°$ to the radial direction; without the vicalloy rotor, this would approach zero. A qualitative explanation of the phenomenon is given below.

Angle of inclination of the flux density

The angle of entry, α, on the pole axis, point E in Fig.C3.11(a) and (b), is considered. The excitation 0.3 A is taken as representative and the corresponding demagnetisation portion of the B-H curve for vicalloy is shown in Fig.C3.12.

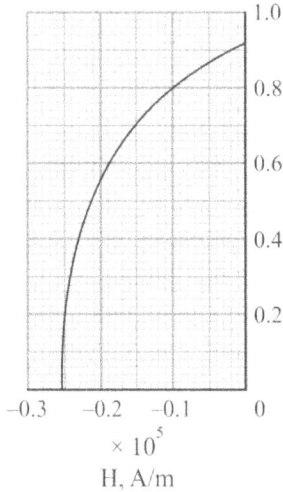

Fig.C3.12 : Demagnetisation curve for vicalloy

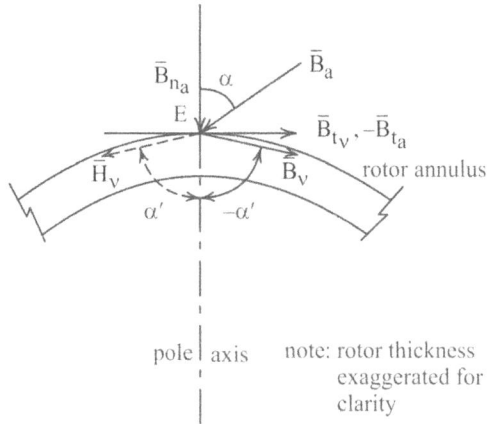

Fig.C3.13 : B_θ and H_θ at point E at 0.3 A excitation

As shown later in Chapter 5, for this part of the rotor B_θ and H_θ are oppositely directed so that the vectors are disposed as shown in Fig.C3.13. Since the flux density in the airgap is small compared to that in the vicalloy annulus, it is assumed that the magnitude of the average B_θ in vicalloy is obtained from the B-H curve using the peripheral component of the airgap magnetising force, and that the angle α' is deduced from B_θ and normal component of the airgap flux density. Hence

in vicalloy
$$\tan\left(-\alpha'\right) = -\frac{B_{t_v}}{B_{n_a}}$$

in airgap
$$\tan\alpha = \frac{B_{t_a}}{B_{n_a}}$$

or
$$\frac{\tan\alpha}{\tan\left(-\alpha'\right)} = \frac{B_{t_a}}{B_{t_v}} \quad \text{(for the same values of } H_t\text{)}$$

If H_t is taken to be 10^{-4} A/m then $B_{t_a} = 1.257 \times 10^{-2}$ and $B_{t_v} = -0.8$ T.

giving
$$\alpha = \arctan\left[\frac{B_{t_a}}{B_{t_v}}\tan(-\alpha')\right] \qquad (C3.1)$$

Substituting in eqn.(C3.1)

$$\alpha = \arctan \left[\frac{\tan(-\alpha')}{63.7} \right] \tag{C3.2}$$

With the assumption that the flux density in the vicalloy annulus is predominantly peripheral, a value of α' would be about $89°$. Accordingly, α would then be about $45°$ or less. A similar situation would obtain at other excitations because of similarity of demagnetisation curves.

This explanation is possible because the value of H_t in the airgap is approximately the same as the average value of H_t in the vicalloy annulus under the pole. It cannot be extended to the inner surface since the approximation is not valid[9].

Hysteresis Effect

When considered reflected in terms of relative difference in heights of peak values of flux density values in the airgap at the two pole tips due to spatial hysteresis, being considerable more at 0.3 A excitation, the effect is not so pronounced on the inner surface. The large peripheral component of flux density results in the 'butterfly-shaped' patterns in the polar plots [see Figs. C3.10 and C3.11].

An advantage of the plots is in indicating the physical location of the 'equivalent' poles in the airgap and arbor region. Considering the direction of the flux density, it would appear that the equivalent pole axes have "shifted" towards the *leading* pole tips during the operation, also pointed out by the B_θ waveforms.

Comparing the measured and computed waveforms of flux density for air rotor with those for vicalloy rotor points to two significant changes during the machine operation:

- concentration of flux in the airgap near the lagging pole tip

- existence of noticeable peripheral component of flux density under the pole

The material outcome of these changes is observed in the form of developed torque in the machine, thus suggesting a relationship between the two effects.

---

[9]This also brings out the importance of use of a thin annulus for the rotor; also reflected in the magnetic measurements on vicalloy reported in Appendix I.

4 : Qualitative Theory of Operation

4

Qualitative Theory of Operation[*]

The foregoing detailed discussion of radial and peripheral flux density distribution in the airgap and arbor regions of the experimental machine can be related to the developed torque in the three identified stages as indicated in the measured torque-excitation curve. In terms of excitation currents, these stages or regions can be identified as

- region I, low excitation: 0 – 0.15 A
- region II, 'normal' excitation: 0.15 – 0.25 or 0.3 A
- region III, higher excitation: 0.4 – 1.0 A

Torque at Low Excitation Currents

The torque equivalent to the alternating hysteresis loss in this region corresponds to the initial part of the magnetisation or loss/excitation curve, relatable analytically to the magnetising force in the rotor. This is because the hysteresis loops (corresponding to any excitation) can be represented in 'parabolic' form[42] to a good approximation and their area calculated mathematically. See Fig.C4.1.

Referring to the figure, the B-H curve for vicalloy (reproduced from Appendix I) can be approximated up to a certain point P by the relationship

$$B = \mu_i H + v H^2 \qquad\qquad (C4.1)$$

where μ_i and v are constants. When evaluated for the present case, these are found to be

$$\mu_i = -0.02 \times 10^{-4} \text{ H/m}$$

and
$$v = 1.24 \times 10^{-9} \text{ H/A}$$

for an upper limit of $H = 0.15 \times 10^5$ A/m. The complete 'hysteresis' loop for the approximation is shown in Fig.C4.2, following page.

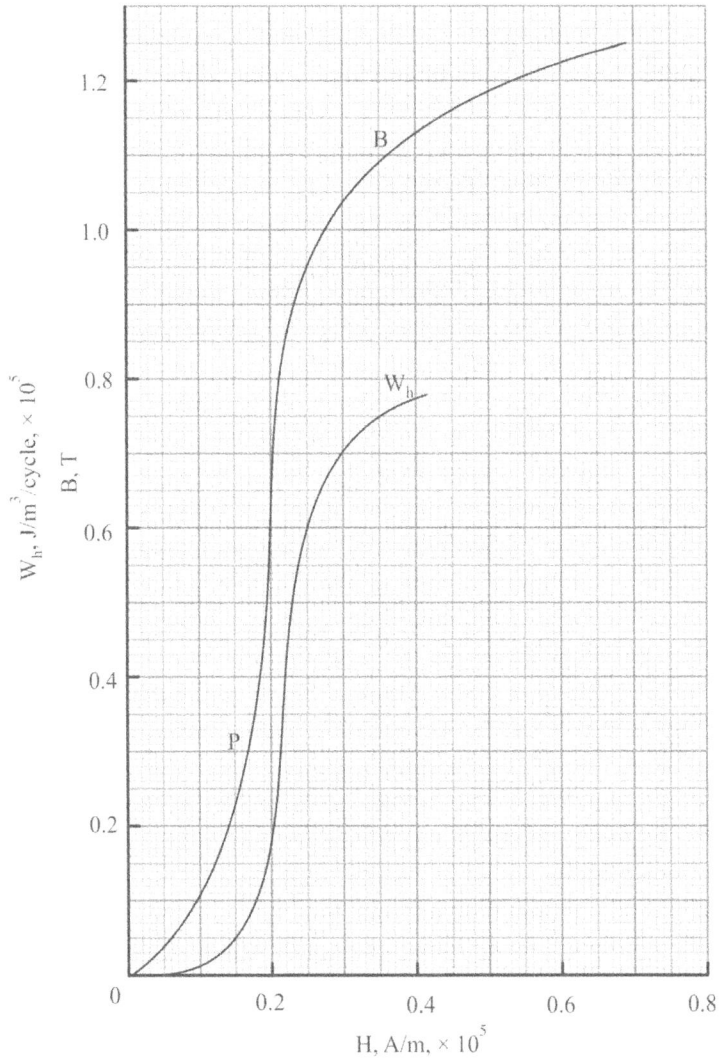

Fig.C4.1 : Magnetisation curve of vicalloy and hysteresis loss

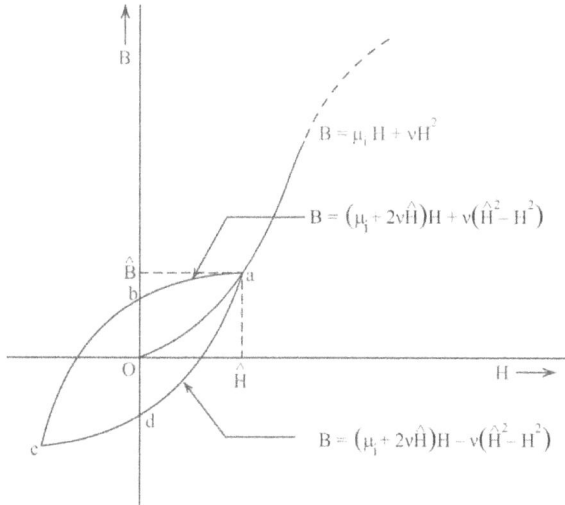

Fig.C4.2 : Hysteresis loop for parabolic approximation of B-H curve

The equation of the loop a b c d a, marked in the figure, is given by[18]

$$B = \left(\mu_i + v\hat{H}\right)H \pm \frac{v}{2}\left(\hat{H}^2 - H^2\right) \qquad \text{(C4.2)}$$

and the area of the closed loop, representing alternating hysteresis loss/m^3/cycle is

$$W_h = \oint B\,dH = 2 \times \int_{-H}^{+H} B\,dH \qquad \text{(C4.3)}$$

Substituting for B from eqn.(C4.2) and simplifying

$$W_h = 4/3\, v\, \hat{H}^3 \qquad \text{(C4.4)}$$

This shows that the curve of alternating hysteresis loss follows a 'cubic' fit or law with regard to applied magnetising force. For the value of constant v evaluated above, W_h is expressed in J/m^3/cycle when $\hat{H}$ is in A/m. Thus

$$W_h = (4/3) \times 1.24 \times 10^{-9}\, \hat{H}^3$$

or

$$W_h = 1.66 \times 10^{-9}\, \hat{H}^3 \text{ J/m}^3\text{/cycle} \qquad \text{(C4.5)}$$

The following table compares the values of the loss calculated from eqn.(C4.5) with that from the measured areas of loops for different values of $\hat{H}$.

Table C4.1: Calculated and 'measured' hysteresis loss

$\hat{H}$, A/m $\times 10^5$	0.1	0.15	0.2
W_h, J/m^3/cycle $\times 10^5$			
(a) using eqn. (C.4.5)	0.01666	0.056	0.133
(b) assessed from hysteresis loop areas	0.015	0.05	0.3

The expression for the loss, W_h, can be converted into developed torque[1]. Thus, if v is the volume of the rotor annulus and p the number of pole-pairs, the torque in Nm is given by

$$T = \frac{p \times W_h}{2\pi} \qquad (C4.6)$$

In the experimental machine, $p = 1$ and $v = 2.67 \times 10^{-6}$ m^3. Hence

$$T = \frac{1 \times 2.67 \times 10^{-6} \times 1.66 \times 10^{-9}}{2\pi} \times \hat{H}^3$$

$$= 0.71 \times 10^{-15} \times \hat{H}^3 \text{ Nm} \qquad (C4.7)$$

Values of $\hat{H}$ in vicalloy are not directly available for use in eqn.(C4.7) for calculation of torque. However, an approximate form of the equation, relating torque to excitation current, can be worked out on the following basis.

For a given excitation, the maximum value of measured/computed B_θ in the vicalloy rotor is available from which corresponding values of $\hat{H}$ can be obtained using the B-H curve shown in Fig.C4.1. Using this approach, a relationship between $\hat{H}$ and I, the excitation current, is determined to be

$$\hat{H} = 2.24 \times I^{1.14}$$

and this when substituted in eqn.(C4.7) yields

$$T = 8.1 \times I^{3.42} \text{ Nm} \qquad (C4.8)$$

The calculated torque values for excitation up to 0.15 A using eqn.(C4.8) are compared with the measured values in Table C4.2.

Table C4.2: Calculated and measured torque at low excitations

Excitation current, I, A	0.05	0.1	0.15
Torque calculated using eqn. (C4.8), Nm	0.000275	0.0031	0.0124
Measured torque, Nm	-	0.004	0.012

It is observed that whilst the two values are within 4.0% at 0.15 excitation, the calculated value for 0.1 A excitation is 22.5% lower than the measured value. Although this discrepancy may be due to experimental error to some extent since accurate measurement of torque at very low currents would not be possible, there are other factors to account for it as discussed later.

Torque in the Normal Excitation Range

The torque in this region increases rapidly in accordance with the steeply rising alternating-hysteresis loss curve (see Fig.C4.1) and the peripheral flux density, B_θ, variation in the rotor annulus; the spatial variation of average B_θ for varying excitation is given in Fig.C4.3.

---

[1]See Chapter V.

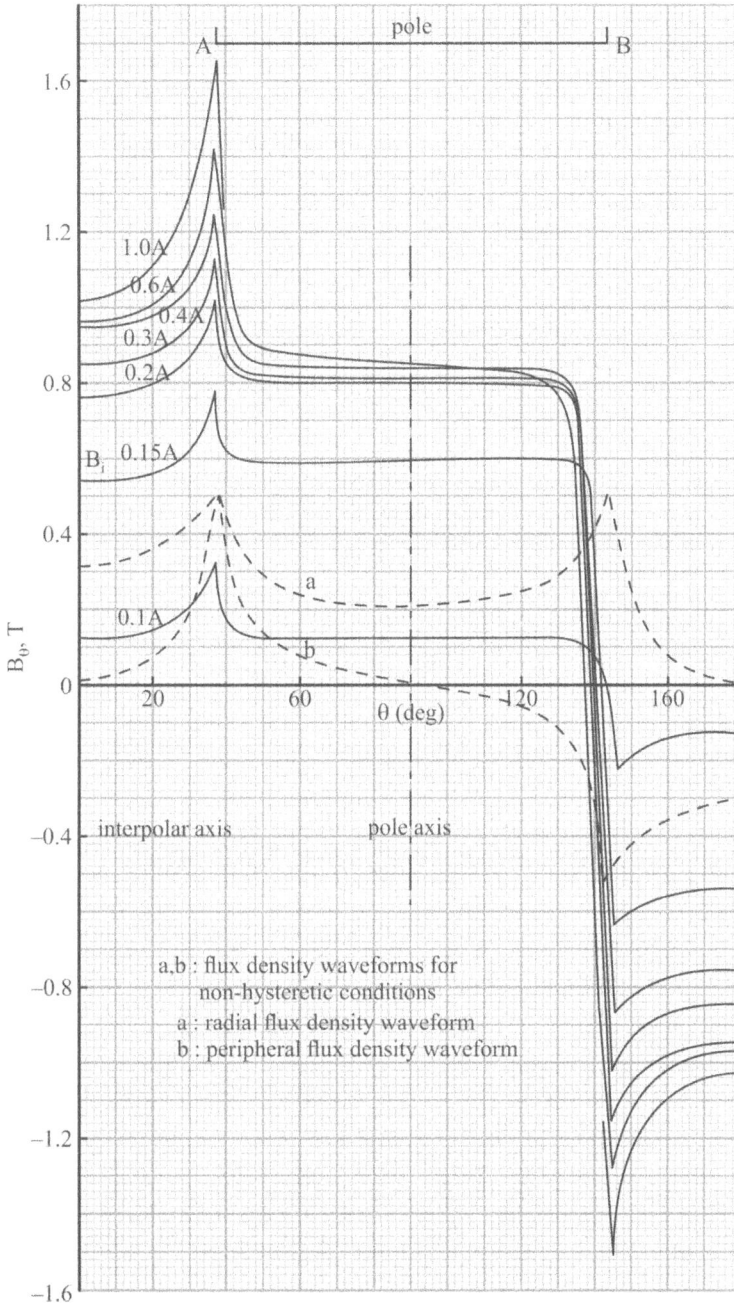

Fig.C4.3 : Peripheral flux density variation with excitation

A relationship between the torque and excitation on the lines as for region I cannot be derived since the hysteresis loops are indescribable in true sense by a single mathematical expression.

Referring to Fig.C4.3, the B_θ variation is seen to be uniform under the pole surface and interpolar region, increasing rapidly in magnitude for excitations from 0.1 A to 0.3 A. This 'growth' of B_θ apparently follows the same trend as the developed torque. Fig.C4.4 shows the plots of the values of B_θ on the interpolar and pole axis, B_i and B_p, respectively, against excitation whilst a plot of B_p against developed torque is given in Fig.C4.5 which shows that for the current range 0.15 A – 0.25 A, the torque is proportional to B_p.

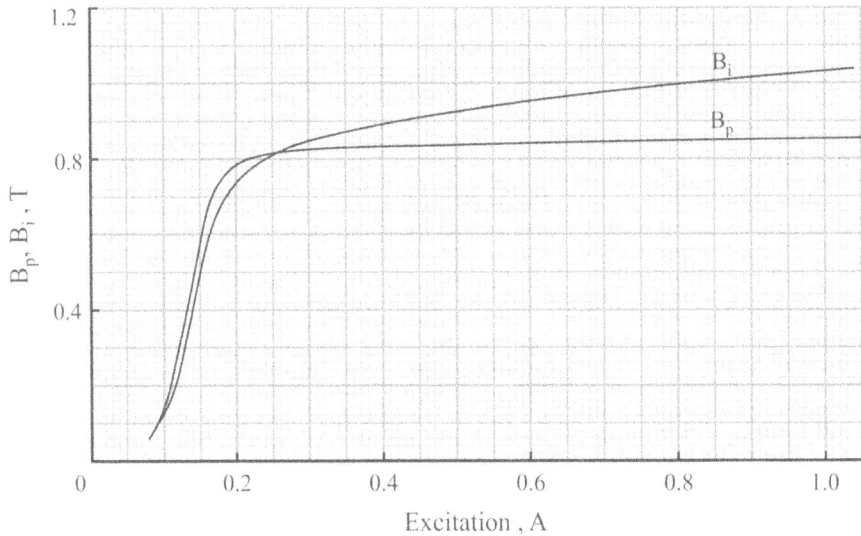

Fig.C4.4 : Variation of B_i and B_p with excitation

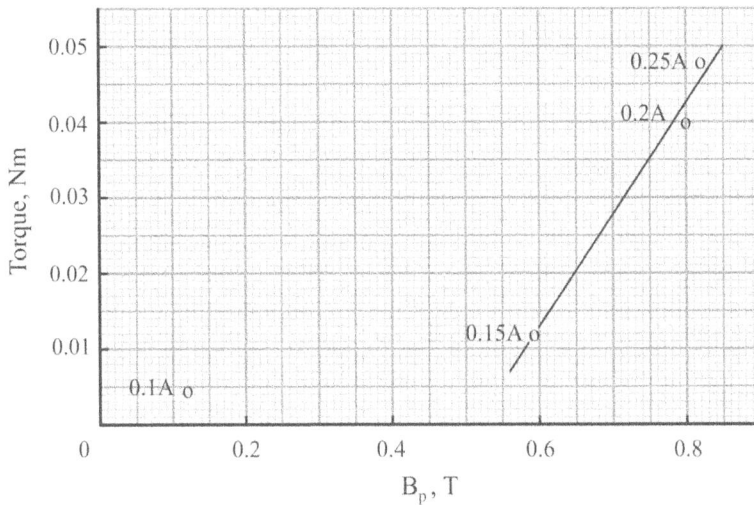

Fig.C4.5 : Variation of torque with flux density, B_p

Assuming that the developed torque is dependent on the 'average' flux density under the pole surface, B_p, the form of a mathematical relationship between B_p and excitation, I, will also govern the developed torque in region II.

This torque expression is obtained as

$$T = 0.29 \times I^{0.62} - 0.077 \text{ Nm} \qquad (C4.9)$$

in the current range 0.15 A $< I <$ 0.25 A. The computed torque curve using eqn.(C4.9), the calculated curve for region I and measured curve up to 0.3 A excitation are shown in Fig.C4.6. It is seen that at the common excitation of 0.15 A, the torque given by eqns.(C4.8) and (C4.9) is the same and within 4% of the measured value. The former relationship, however, is simpler to apply and related to alternating loss more realistically.

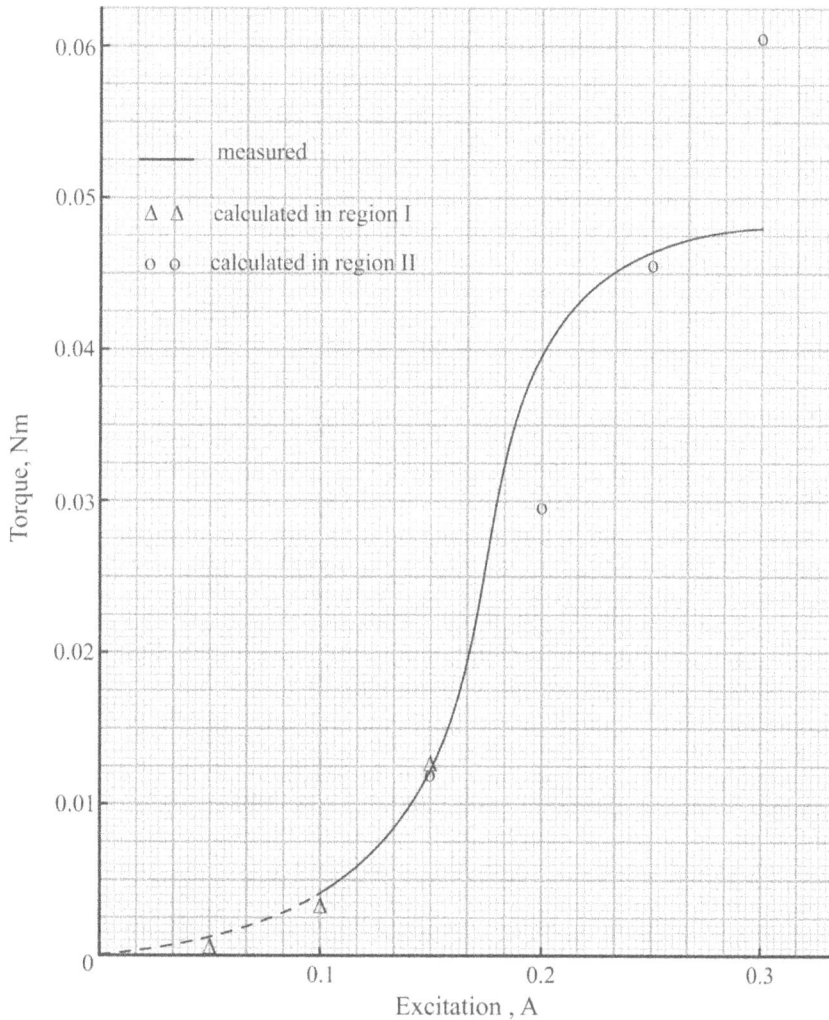

Fig.C4.6 : Torque vs. excitation in regions I and II by calculation and measurement

A general empirical torque expression

The torque-excitation curve in regions I and II is a smooth curve and it is possible to derive a general, 'empirical' torque expression for the complete range of 0 to 0.25 A.

Assume that the curve follows the expression

$$T = k \ I^n \tag{C4.10}$$

where k is a constant and n a positive number.

A plot of T and I (on log-log paper) is shown in Fig.C4.7 whence the values of k and n are found to be

$$K = 8.0 \ \text{and} \ n = 3.3$$

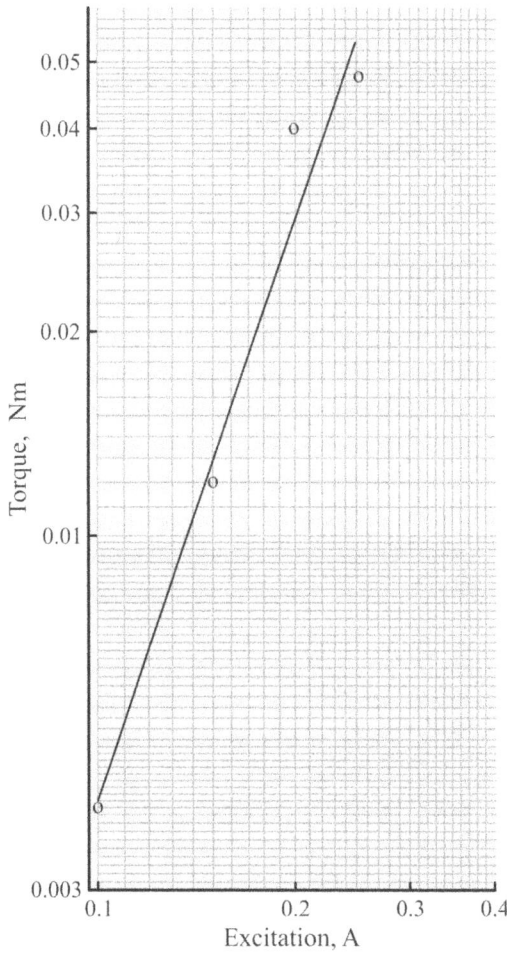

Fig.C4.7 : Log-log plot of torque vs. excitation

Therefore the general torque expression is given by

$$T = 8.0 \times I^{3.3}$$

and has a close resemblance with eqn.(C4.8) for region I.

Torque at Higher Excitations

The developed torque in region III shows a steady *decrease* above an excitation of (about) 0.3 A^2. The reduction of torque at higher currents is a special characteristic of the experimental machine. Theoretically, the variation of developed torque at increasing magnetising force or excitations should follow the alternating hysteresis loss, W_{h_a}, with mmf[39]. This is depicted in Fig.C4.8 which shows that the loss goes on increasing beyond the initial bend a, though at a slower pace, until at point b beyond which it tends to level off.

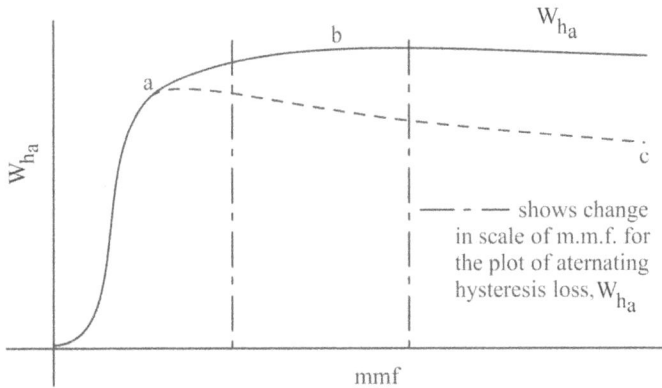

Fig.C4.8 : Alternating hysteresis loss vs. mmf

That the developed torque follows an opposite trend at higher excitation or mmf, indicated by (dotted) line a c, is due to changes occurring in the flux distribution in the machine as discussed below.

Rotor Magnetisation

Peripheral flux density waveforms in the rotor

A noteworthy feature of the magnetism in the rotor is its effect on the point of reversal in the waveform of average peripheral flus density[3]. One of these waveforms from the set

---

[2]This makes the torque at 0.3 A a maximum, resulting in a hump between 0.25 A and 0.4 A. Undue prominence need not be given to the existence of a maximum torque; it is not a feature of, say, every hysteresis motor design although there may always be a noticeable change of slope corresponding to the change of slope in the alternating hysteresis loss curve,

[3]The word average implies derived flux density obtained from the single turn search coil wound round the annulus cross section.

shown in Fig.C4.3 for the excitation 0.3 A is reproduced in Fig.C4.9(a) and (b) together with that computed for the external rotor surface in the airgap.

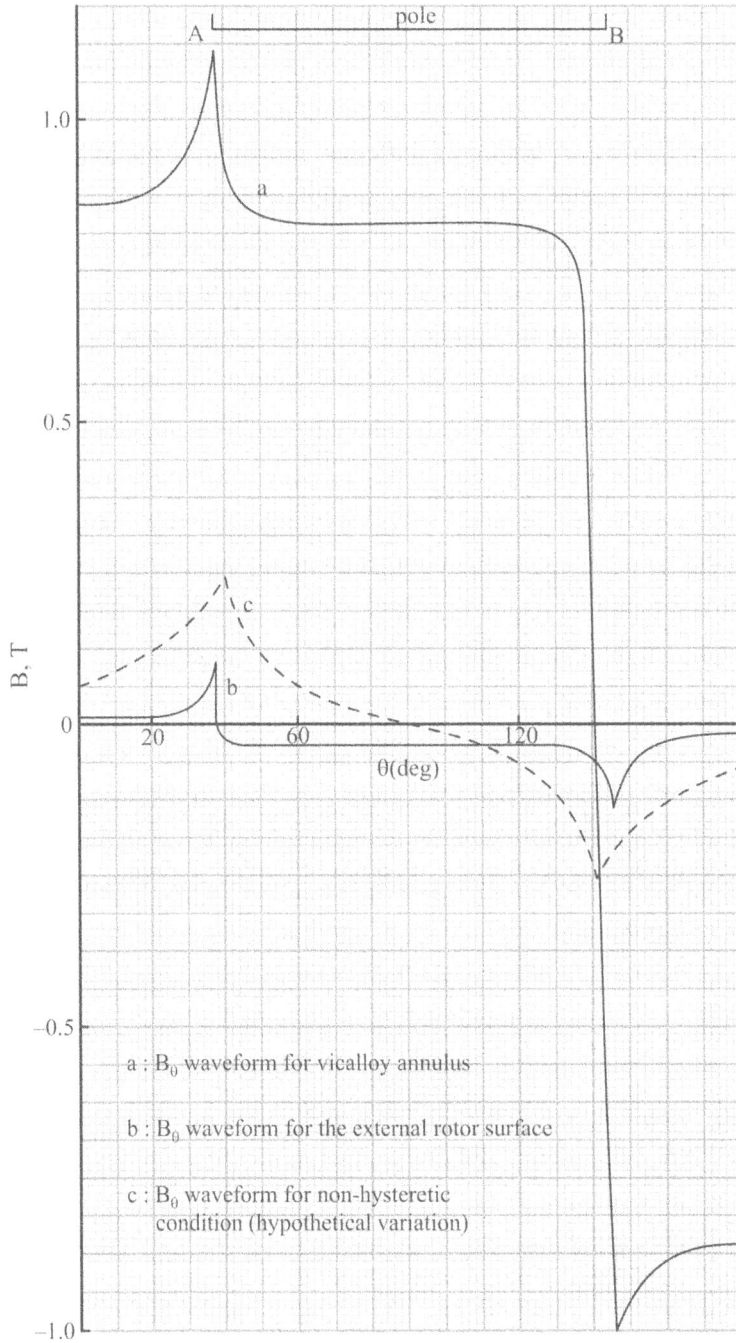

a : B_{θ} waveform for vicalloy annulus

b : B_{θ} waveform for the external rotor surface

c : B_{θ} waveform for non-hysteretic
    condition (hypothetical variation)

(a)

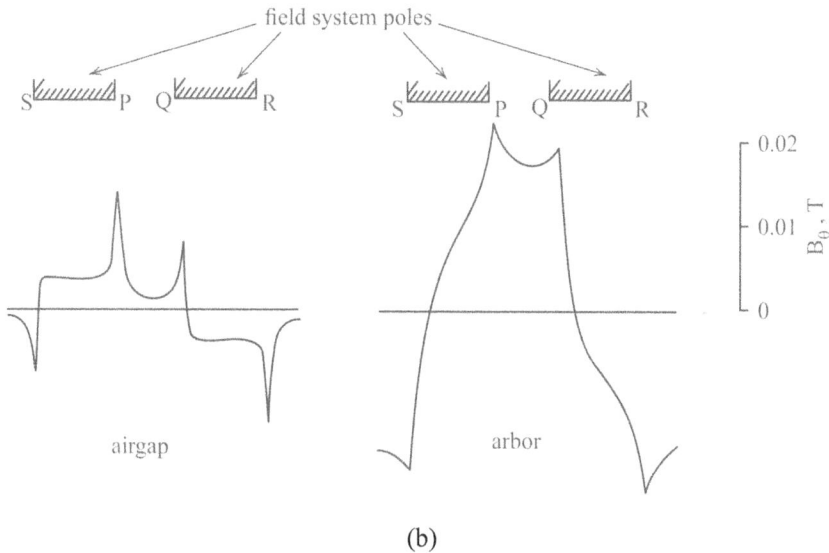

(b)

Fig.C4.9 : Peripheral flux density variations in the rotor at 0.3 A excitation

It is seen that whilst the reversal in the B_θ waveform on the surface (curve b in Fig.C4.9(a)) occurs near the *leading* pole tip A, that in the waveform for the annulus (curve a) takes place opposite the *lagging* pole tip B. This means that the flux being predominantly peripheral within the annulus owing to its small thinness divides at a point opposite the lagging pole tip. Note that in the non-hysteretic rotor the flux divides equally opposite the pole centre, curve c, in Fig.C4.9(a). There is no evidence from the waveforms to show that the geometrical pole centre has any effect on the flux distribution or that more than two reversals occur in the flux direction within the rotor per cycle. This shows that the flux leaving the lagging pole tip which is more than that from the leading pole tip [see Figs.C3.11(a) and (b), Chapter 3] dominates the flux pattern in the rotor, leading to the division of this flux taking place opposite the lagging pole tip in the annulus.

These waveforms illustrate yet another effect of (spatial) hysteresis at *all* values of excitation. In the absence of hysteresis, the reversal would occur at the pole centres. Also, hysteresis affects the B_θ waveform in the arbor; as the excitation is progressively increased, the points at which the waveform reverses in the sign move from the pole-centres, reach a maximum near to the leading pole-tips, and then return to the pole-centres.

The condition depicted in Fig.C4.9 is presented qualitatively in Fig.C4.10.

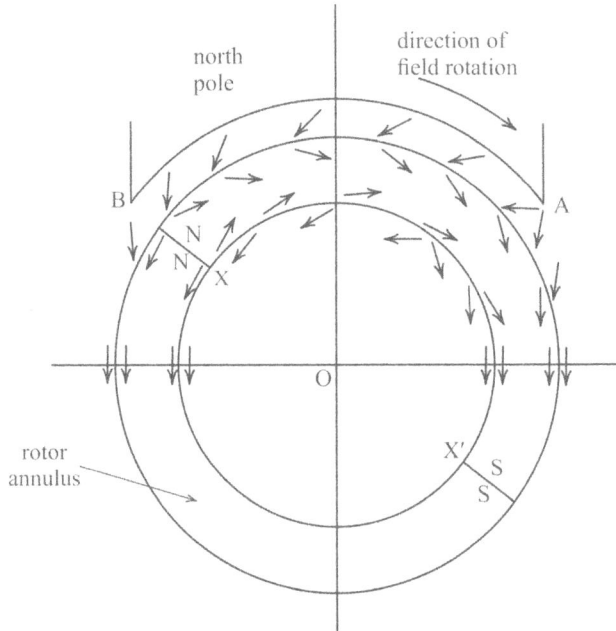

Fig.C4.10 : Magnetisation of vicalloy rotor

It is evident that the rotor behaves as if it comprised two permanently magnetised semi-annular cylinders butted together at the interfaces X, X′. The two cylinders exhibit the same polarities at the interfaces, NN and SS, as shown. Clearly, the existence of these 'poles' opposite the pole tips during the machine operation is the effect of spatial hysteresis in the rotor since in the absence of such an effect the 'poles' would occur under the field-system pole centres[4].

Equivalent Vacuo Model for the Rotor

The above phenomenon observed in the rotor can be explained in terms of characteristic aspects of permanent magnetism. To apply this concept, a magnetisation vector or magnetic (dipole) polarisation, $\overline{M}$, is introduced to represent the rotor magnetisation and an analogy in terms of fictitious magnetic surface charges on interfaces of Fig.C4.10 whilst proposing an equivalent vacuo model shown in Fig.C4.11.

---

[4] In the figures depicting rotor annulus, the radial thickness, as also the airgap length, such as in Fig.C4.10 and more later, are exaggerated for clarity.

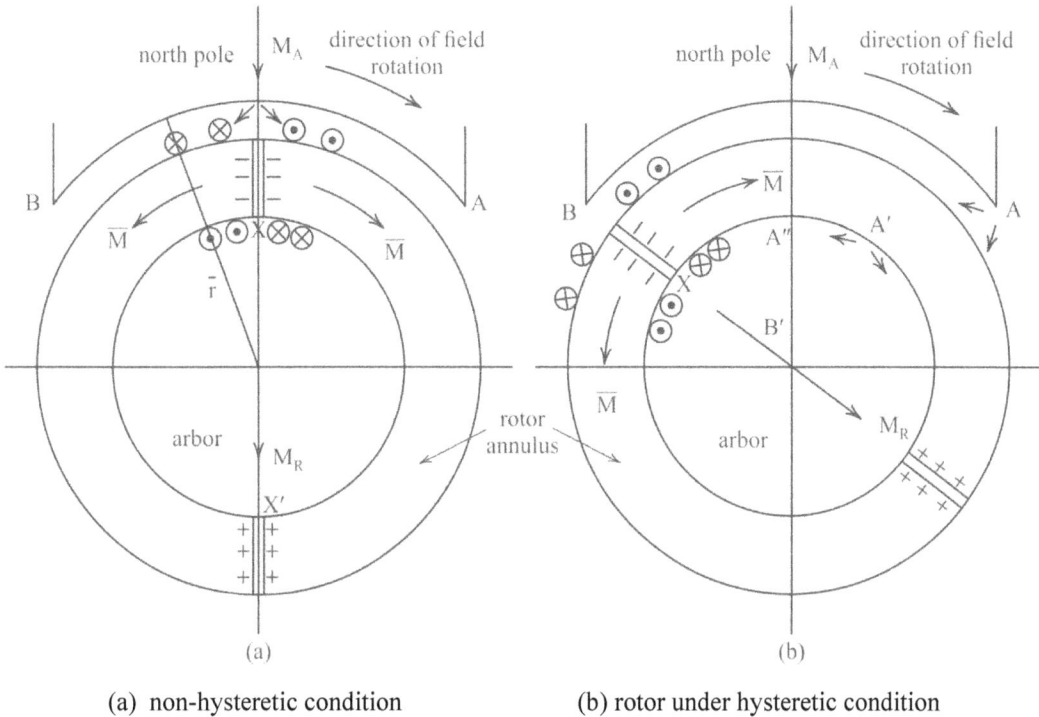

(a)   non-hysteretic condition   (b) rotor under hysteretic condition

Fig.C4.11 : Vacuo model of the rotor and magnetic charge distribution

The surface charges are distributed as shown[5], the direction of 'rotor magnetisation' vector being indicated by arrows within the rotor annulus. Fig.C4.11(a) illustrates the non-hysteretic condition when the flux from the pole (assumed to be a north pole) divides equally at the centre both in the airgap as well as the annulus. Treating the rotor as a single magnetised body, this is equivalent to a 'resultant' rotor mmf, M_R, in space phase with the applied mmf, M_A, since $\overline{M}$ is equal but oppositely directed at X. This condition also exists when the field system is held stationary and the rotor magnetised by the mmf M_A.

Fig.C4.11(b) shows the condition when the rotor mmf lags the applied mmf, identified as 'the hysteretic effect', and the division of flux within the annulus occurs opposite the lagging pole tip B. The distribution of $\overline{M}$, as seen from the field system, is modified accordingly such that within the region AB of the airgap and $A'B'$ on the inside of the rotor, the tangential component of $\overline{H}$ is *oppositely directed* to $\overline{B}$ within the vicalloy annulus.

---

[5]Ideal conditions are assumed, that is uniform magnetisation of the annulus such that $\overline{\nabla} \cdot \overline{M} = 0$.

This magnetic condition is analogous to the conditions prevailing in a (permanent) bar magnet where the lines of $\overline{B}$ coincide with $\overline{H}$ outside the magnet, but are oppositely directed inside as illustrated in Fig.C4.12(a).

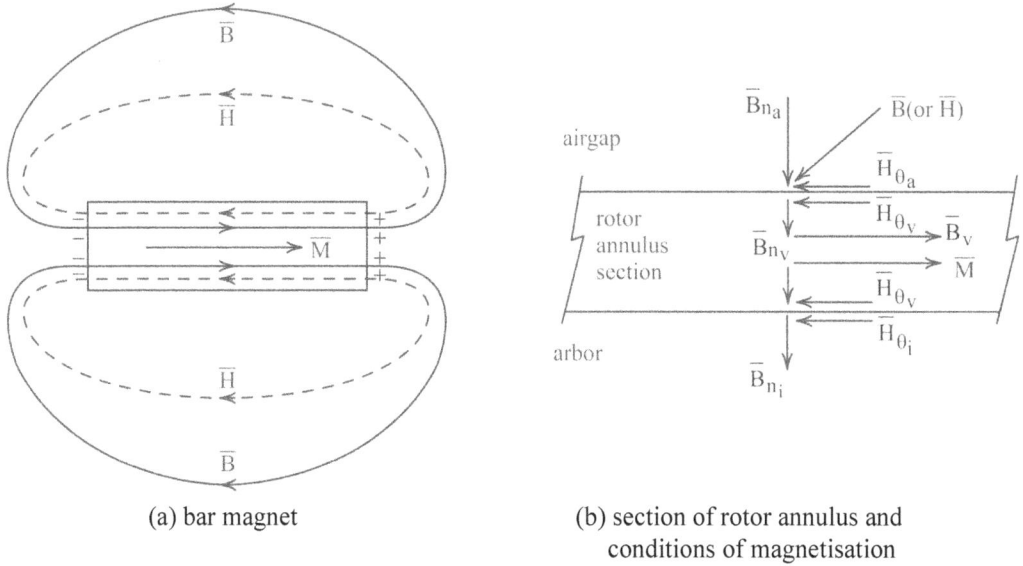

(a) bar magnet   (b) section of rotor annulus and
                                      conditions of magnetisation

Fig.C4.12 : Magnetisation of a bar magnet and rotor annulus

The diagram of Fig.C4.12(b) represents a section of the rotor annulus in the common region   A′ B′, [Fig.C4.11(b)], and shows the field components for the airgap, the vicalloy rotor and the arbor region. It is observed that in spite of the peripheral flux density, B'_v, in vicalloy being directed opposite to $\overline{H}$ on either side, the boundary conditions are satisfied; that is, the normal components of $\overline{B}$ and tangential components of $\overline{H}$ are continuous[6]. It follows, therefore, that under the influence of strong rotor magnetisation, the peripheral flux density in the annulus is largely due to $\overline{M}$, being governed by the relation

$$\overline{B} = \mu_o (\overline{M} + \overline{H}) \qquad \overline{M} \gg \overline{H}$$

In the interpolar region, however, the 'original' field due to the field system is intensely peripheral. This means that $\overline{H}$ in the air region as well as $\overline{M}$ and the

---

[6]The two well-known boundary conditions are derived from
(a) normal components

$\oint\!\!\!\oint \overline{B} \cdot \overline{i}_n \, ds = 0$ ∴ $B_{1_n} \delta s + (-B_{2_n}) \delta s = 0$ on a 'small' surface, giving $B_{1_n} = B_{2_n}$

(b) tangential components ($J_s = 0$)

$\oint \overline{H} \cdot d\overline{l} = 0$ ∴ $H_{1_t} \delta l + (-H_{2_t}) \delta l = 0$ enclosing a 'small' contour, giving $H_{1_t} = H_{2_t}$

tangential component of $\overline{H}$ in the vicalloy annulus have the same direction. Fig.C4.13 summarises the directions of the field vectors for the different regions of the rotor.

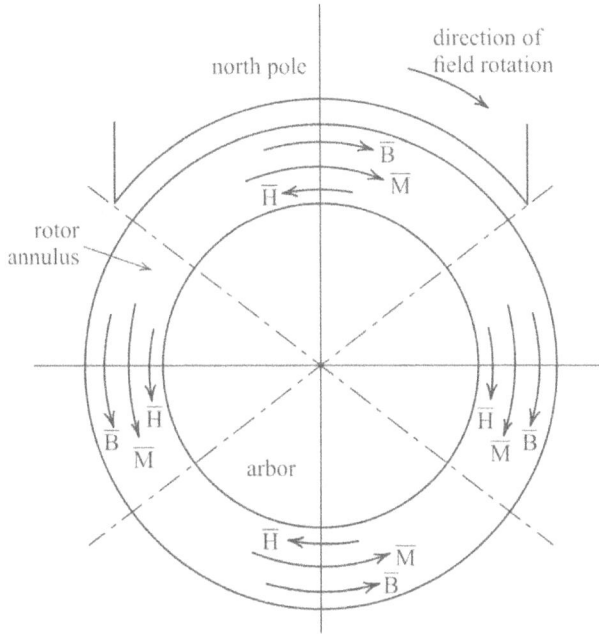

Fig.C4.13 : Field vectors for different regions of the rotor

The foregoing considerations point to a unique aspect of flux density distribution around the periphery of the rotor annulus. Since the flux density in the sections under the poles is predominantly due to $\overline{M}$, it is representative of the intensity of magnetisation, $\overline{M}_0 (= \overline{B} - \mu_0 \overline{H})$, for vicalloy whilst that in the interpolar region is the 'true' flux density given by

$$\overline{B} = \mu \overline{H}, \quad \mu = \mu_0 \mu_r$$

This interpretation is in close agreement with the measured waveform of average flux density in the annulus at increasing excitation as revealed in Fig.C4.4 in which the graph of B_p, equivalent to M_0 according to the present interpretation, indicates intense saturation at higher excitations, not observed in the plot of B_i.

Rotor Field Distribution and Cycle of Magnetisation

Further corroboration of the magnetic phenomenon in the rotor discussed above is provided by the traversal of the magnetisation cycle in space.

The cyclic variation of a magnetic material as shown in Fig.C4.14 has four changeover points, disregarding the shape of the loop at this stage.

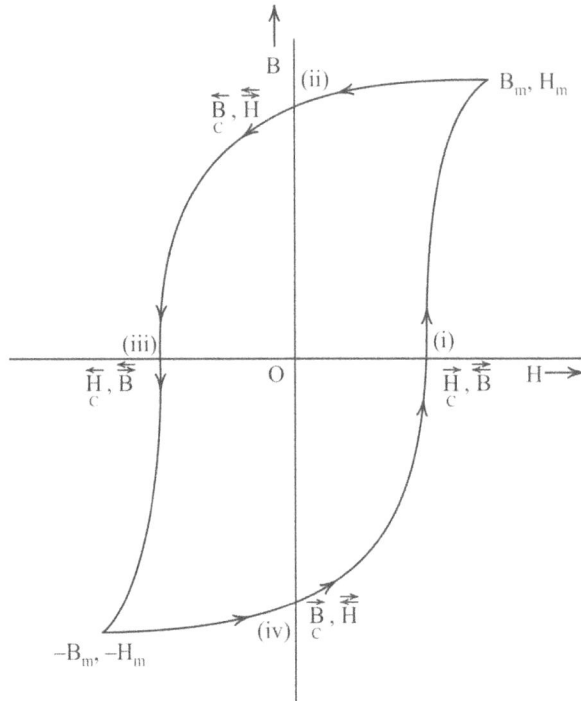

Fig.C4.14 : Cyclic variation in a magnetic material

(i) $\vec{H}, \overset{\leftrightarrows}{B}$: Near constant H in a '+ve' direction, reversal of B from '-ve' to

'+ve' direction

(ii) $\vec{B}, \overset{\leftrightarrows}{H}$: Near constant B in a '+ve' direction, reversal of H from '+ve' to

'-ve' direction

(iii) $\overset{\leftarrow}{H}, \overset{\leftrightarrows}{B}$: Near constant H in a '-ve' direction, reversal of B from '+ve' to

'-ve' direction

(iv) $\overset{\leftarrow}{B}, \overset{\rightleftarrows}{H}$: Near constant B in a '-ve' direction, reversal of H from '-ve' to

'+ve' direction

The four key points define completely the B-H loop and must occur in the same cyclic
order.

The excursion of the hysteresis loop in a similar manner occurs in the rotor annulus for all points around the periphery at any instant or for a given point in space with varying time. The magnetic conditions for the rotor in space are illustrated in Fig.C4.15(a) with $\bar{B}$ and $\bar{H}$ assumed to have constant magnitudes.

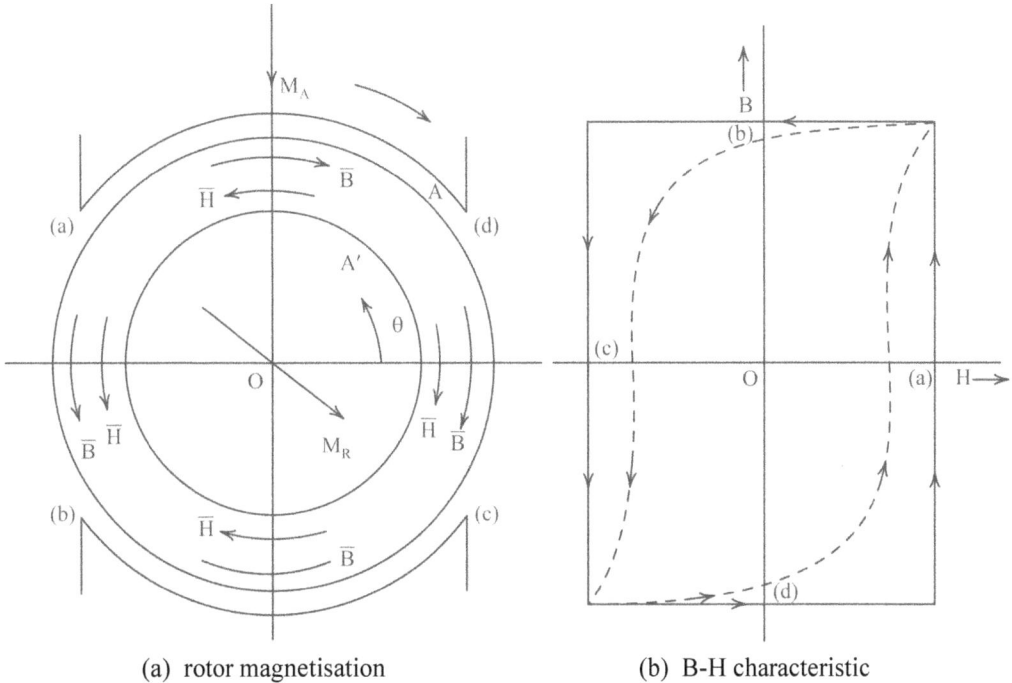

(a) rotor magnetisation   (b) B-H characteristic

Fig.C4.15 : Cyclic magnetisation of rotor annulus

The corresponding magnetisation loop is shown in Fig.C4.15(b) which is of rectangular shape because of assumed constant values of B and H. The four 'operating points' are the FOUR corner points and the loop is traversed as a series of 'jumps' between these points. In a physical model, the value of B and H cannot be constant and the loop would be replaced by the usual form, shown dotted in Fig.C4.15(b). Note that in the actual operation of the machine although the sequence represented by Fig.C4.15 still holds, the hysteresis loop would be modified and must include recoil loops owing to the shape of flux density waveforms. No experimental verification of this phenomenon is possible since the magnetic excitation waveform for vicalloy cannot be obtained. Nevertheless it is a reasonable assumption and can be used to deduce the average magnetising force variation if approximate recoil loops are sketched in. Assuming that the B_θ variation in vicalloy represents the resultant flux density at any point, neglecting any radial component, then the H_θ variation would be as shown in Fig.C4.16 for an excitation of 0.3 A, the waveform seen to be containing appreciable peaks.

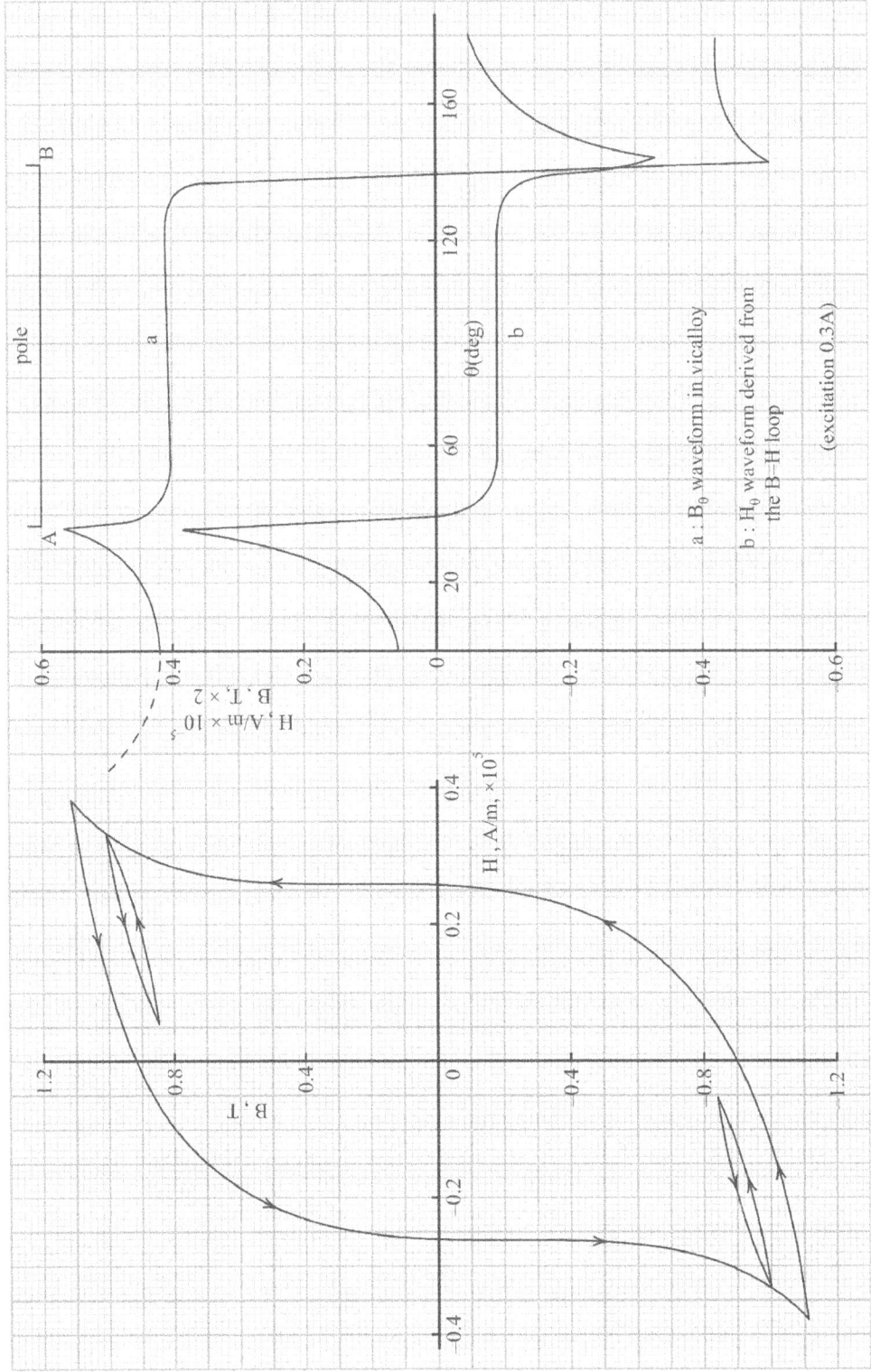

Fig.C4.16 : Average B_θ variation and derived H_θ variation in rotor annulus

The values of H_θ as obtained from the hysteresis loop are plotted in Fig.C4.17, together with corresponding B_θ in vicalloy and H_θ on the external rotor surface (see Fig.C3.5, Chapter 3) showing a close agreement[7].

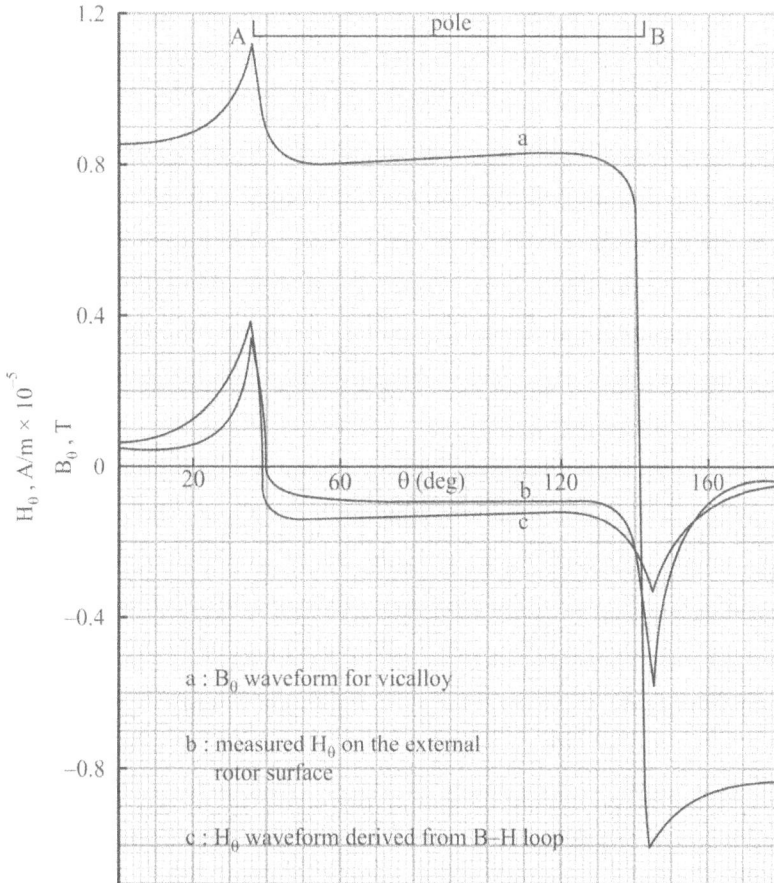

Fig.C4.17 : Derived H_θ and measured B_θ waveforms in vicalloy rotor

Minimum Energy Conditions in the Rotor

Under the assumption of uniform rotor magnetisation, the 'effective' rotor mmf, M_R, was taken to be in the direction BB' on the basis of the flux dividing in the *annulus* opposite to pole tip B as depicted in Fig.C4.11. It may be argued that this mmf could instead lie in the direction AA as shown in Fig.C4.18 since the flux in the *airgap* divides at pole tip A and not B during the machine operation [cf. Fig.C4.11(b), indicated by 'small' arrows].

---

[7]The recoil loops occurring in the magnetisation cycle during the machine operation might contribute to the developed torque; however, considering their relatively small size or area, their contribution would be negligible.

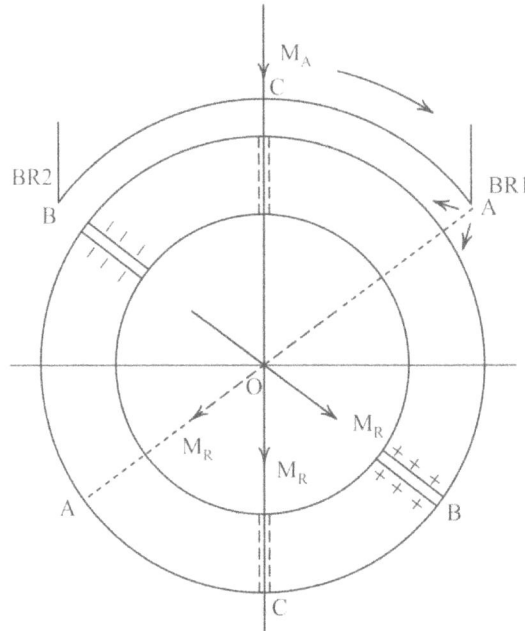

Fig.C4.18 : Alternative direction of resultant magnetisation

However, a reference to the polar plots for the airgap, Fig.C3.11(a) and (b), Chapter 3, shows that the resultant flux density at pole tip B, BR2, is far greater than that at pole tip A, BR1. Since the entire rotor is being treated as a single magnetised body, its effective mmf is directed such that the total energy of the system, that is the rotor and the field system, is always a minimum. Compliance with this condition is dictated in terms of BR2 being greater than BR1 and hence M_R directed along BB. Accordingly, if the field system were (held) stationary and the rotor were free to rotate, the minimum-energy requirement would cause the rotor to revert to the conditions of Fig.C4.11(a) when M_R would be directed along CC as indicated in Fig.C4.18[8].

The only limitation of the foregoing discussion in practice can be that the rotor magnetisation is not uniform along the entire periphery of the rotor, that is $\bar{\nabla} \times \bar{M} \neq 0$, esp. at excitations outside the range of 0.15 A to 0.4 A as reflected by appreciable peaks occurring in the B_θ waveform for vicalloy near the pole tips.

Magnetic Flux Density Patterns on the Internal Rotor Surface

As seen from the B_θ waveforms of Fig.C3.7, Chapter 3 and the resultant flux density plots, the flux on the inner rotor surface divides at a point such as A', Fig.C4.11(b), under a combined effect of the flux leaving the inner rotor surface and that crossing the annulus

---

[8]This condition is analogous to synchronous operation of a conventional hysteresis motor as represented by P in the torque-slip 'curve' of Fig.C2.3, Chapter 2.

from the pole tip A. A distinct trend of a path of magnetic flux, on the basis of available waveforms in the airgap and arbor region, is not possible owing to the role played by the rotor magnetisation. From A' to B', Fig.C4.11(b), H_θ and B_θ in vicalloy are oppositely directed, the boundary conditions being satisfied as indicated in Fig.C4.12. The tendency for the flux to bifurcate at A' also accounts for an increase in radial flux density near the leading pole tip rather than the lagging pole tip. If no hysteresis effect were present, the point of flux division must occur at A″ as in Fig.C4.11(b). However, as the flux in the airgap divides at pole tip A under hysteretic condition, the corresponding point on the inner surface is shifted from the pole centre to A', a point intermediate between the two extreme positions. This represents weak hysteresis effect prevailing on the inner rotor surface.

Production of Torque

The concept of rotor magnetisation is an integral feature of the phenomenon of torque production in the experimental machine. However, it is useful to consider first the principles of torque production.

Basic requirements

The simplest physical explanation for the production of driving torque in an electric machine is normally couched in terms of magnetic lines of force in strain in the *airgap*. In many cases this situation is arrived at by the action of the field due to current carrying conductors on the rotatable member on the field originally existing in the airgap.

A similar explanation applies to the experimental machine from the comprehensive study of flux distribution reported in Chapter 3. The angles at which the flux enters at different points on the rotor surface as derived from the airgap flux density distribution are reproduced diagrammatically in Fig.C4.19.

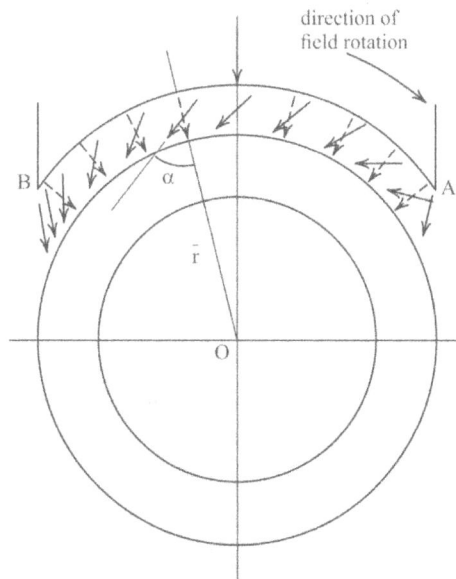

Fig.C4.19 : Strained magnetic field in the airgap of the machine

Shown in the figure are lines of force, indicated by small, 'solid' arrows, *strained due to rotor hysteresis* and the applied magnetic field due to field system. Had there been no spatial hysteresis effect caused by the rotor, as for example at very low excitation currents or with the air rotor, the magnetic field in the airgap would be radially directed as indicated by dotted arrows. Effectively, this means that rotor hysteresis when considered independently provides the 'secondary magnetic flux' that would react with the primary or original field resulting in production of torque according to 'B *I* I' concept.

The inclined field in the airgap therefore emerges with a particular meaning. According to Maxwell's stress consideration, the angle at which the resultant flux enters the rotor surface is a significant factor in deciding the magnitude of the developed torque. For maximum tangential force acting on the rotor surface, this angle should be $45°$ at any point with respect to the radius vector, that is $\alpha = 45°$, Fig.C4.19. On the external rotor surface and under the pole centre, this value of α is very nearly realised; the angle lies between $45°\text{-}40°$ for excitations in the range 0.3 A – 1.0 A, the corresponding Maxwell stress being shown in Fig.C4.20.

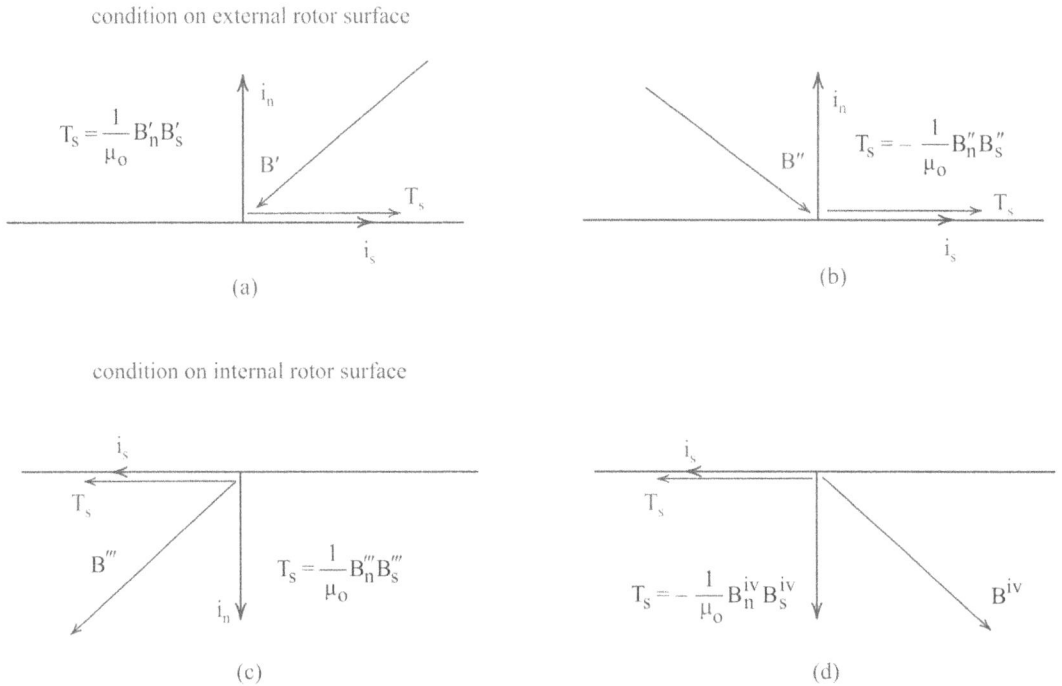

condition on external rotor surface

$$T_s = \frac{1}{\mu_o} B'_n B'_s$$

(a)

$$T_s = -\frac{1}{\mu_o} B''_n B''_s$$

(b)

condition on internal rotor surface

$$T_s = \frac{1}{\mu_o} B'''_n B'''_s$$

(c)

$$T_s = -\frac{1}{\mu_o} B^{iv}_n B^{iv}_s$$

(d)

Fig.C4.20 : Maxwell's stress on the external and internal rotor surface

This is explained as follows.

Maxwell stress $\overline{T} = \overline{H}(\overline{B} \cdot \overline{i}_n) - \frac{1}{2}(\overline{H} \cdot \overline{B})\overline{i}_n$ as depicted in Fig.C4.21.

$$T_n = T \cos 2\theta$$
$$T_s = T \sin 2\theta$$

Case I: surface is on an equipotential, $\theta = 0$ pure tensile stress: $T = \frac{1}{2} B$

Case II: $\bar{H}$, $\bar{B}$ lies in the surface acts as a 'pressure' (opposite to $\bar{i}_n$): $T = \frac{1}{2} B H$

Other values of α are to be found, particularly in the interpolar region where the angle and corresponding stress is negative as shown in Fig.C4.20(b). It follows that the net force is a combination of both positive and negative peripheral surface stresses.

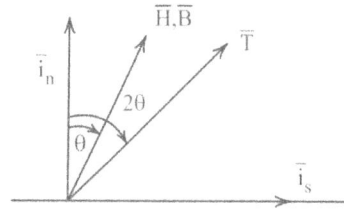

Fig.C4.21 : Schematic of Maxwell stress

On the internal surface, similar angles are to be found although these tend to have a higher value, typically $60°$ tp $70°$ in the excitation range 0.1 A, to 1.0 A, meaning that the magnitude of force, applying Maxwell stress criterion, is smaller. Again, both positive and negative values of α and surface stress are present, the conditions being shown in Fig.C4.20(c) and (d).

The net force on the external rotor surface is in the direction of the rotating field (system) whilst that on the internal surface is oppositely directed and smaller in magnitude; the net torque on the annulus also being in the direction of rotating field.

Clearly, the existence of an inclined magnetic field as the necessary condition for production of torque, requires that a *peripheral* component of $\bar{B}$, B_θ, (or H_θ) *must* exist in the airgap under the pole surface and this should be unidirectional if the maximum possible torque is to be realised at a given excitation. The lagging rotor mmf resulting from the spatial hysteresis effect fulfils this requirement.

Rotor Magnetisation and Production of Torque

Inclined magnetic field in the airgap

The physical picture of radial lines of force in the airgap (shown dotted in Fig.C4.19) becoming inclined under the action of a lagging rotor mmf emerges from the consideration of an 'independent' rotor magnetisation. This property of the rotor was earlier referred to as interface of the semi-annular cylinder being a fictitious 'pole', Fig.C4.10. These poles can be imagined to 'pull' the lines of force in the airgap from their normal direction and result in a flux concentration towards the lagging pole tip, B, the resultant effect being in line with the "minimum-energy-requirement" condition.

An alternative, convincing explanation for the inclined field can be developed by replacing the rotor magnetisation by *surface currents*, analogous to armature reaction in a DC generator.

Alternative vacuo model for rotor magnetisation

To explain the production of torque in terms of rotor magnetisation $\bar{M}$, the concept of equivalent fictitious 'magnetic' surface currents can be introduced; the latter give rise to $\bar{B}$ which when combined with $\bar{M}$ yields $\bar{H}$. This is graphically depicted in Fig.C4.22.

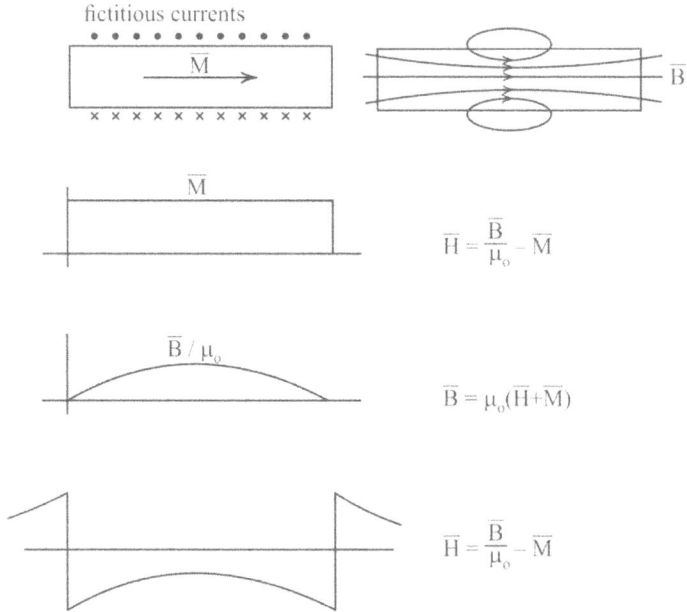

Fig.C4.22 : Fictitious surface currents and magnetic field

Under this assumption, the rotor annulus can be replaced by current elements distributed on the surface according to the relation

$$\overline{J}_s = \overline{H} \times \overline{i}_n \qquad (C4.11)$$

where $\overline{J}_s$ is the surface magnetic current density and $\overline{i}_n$ is a unit vector directed normal to the surface[9]. These current elements are shown as $\otimes$ $\odot$ in Fig.C4.11.

The current elements on the same radius vector such as α' α'' in Fig.C4.23 can be regarded as closed loops/dipoles, each having a dipole moment $\mathscr{M}_\Theta$. The complete rotor annulus thus consists of a very large number of identical dipoles situated along the rotor periphery. Under the action

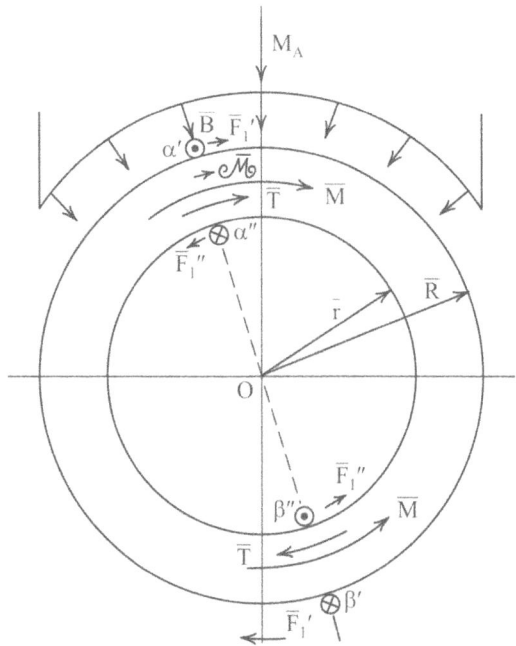

Fig.C4.23 : Current elements along the rotor periphery

[9] Again, uniform magnetisation is assumed such that volume current density $\overline{J}_v = \overline{\nabla} \times \overline{H}$ is zero.

of an applied $\overline{B}$ field, each dipole would experience a torque governed by the expression

$$\overline{T} = {}_e\overline{M}_b \times \overline{H} \tag{C4.12}$$

which will be maximum when ${}_e\overline{M}_b$ and $\overline{B}$ are orthogonal.

Reference Fig.C4.23, the torque $\overline{T}$ is directed to rotate the dipole in a clockwise direction so as to make its plane at right angles to $\overline{B}$. Taking into account the location of the current elements constituting the dipole the above tendency is equivalent to a couple acting on the dipole with the forces $\overline{F}_1'$ and $\overline{F}_1''$ in the directions shown. In terms of the annulus radii $\overline{r}$ and $\overline{R}$, $\overline{F}_1'$ and $\overline{F}_1''$ result in torques on the rotor given by

$$\overline{T}_1 = \overline{F}_1' \times \overline{R} \qquad \text{and} \qquad \overline{T}_2 = \overline{F}_1'' \times \overline{r}$$

with $\overline{T}_1$ opposing $\overline{T}_2$. However, since $\overline{R} > \overline{r}$, the resultant torque on the rotor is seen to be in the clockwise direction under the assumed field distribution of Fig.C4.23 and this is added to the torque arising from the current elements $\beta' \beta''$ at the opposite side of the rotor.

Comment

It follows that for equal surface currents or *uniform magnetisation of the rotor*, the torque developed as a result of dipole moments is a function of rotor radii. In the qualitative theory, with the torque proportional to total hysteresis loss in the rotor active material, the above conclusion points to the requirement of a thick rotor to yield high value of developed torque.

Mathematically

Let N = number of dipoles/unit angle

Therefore, in an angle $d\theta$, number of dipoles = $N\, d\theta$

For constant current in the elements, the magnitude of a dipole moment is

$$ {}_e\overline{M}_b \propto I\,(R - r) \qquad \text{on any radius vector}$$

The torque at the shaft due to dipoles in angle $d\theta$ is

$$ T' \propto I\,(R - r)\, d\theta $$

and the total torque

$$ T \propto \int_0^{2\pi} I(R - r)\, d\theta $$

or proportional to the volume of the rotor annulus.

In practice, this is not realised because the rotor magnetisation is NOT uniform and the "change-over" between the two directions of rotor magnetisation no longer takes place along a radial plane.

Developed Torque in the Higher Excitation Range

The reduction of torque at higher excitations is almost linearly proportional to the current as seen in the graph of Fig.C2.1, Chapter 2. This can be related to the modification in the flux distribution in the machine.

Effect of flux distribution

The concentration of flux at the lagging pole tip in the airgap is the key factor associated with the developed torque. The extent to which this phenomenon affects the measured torque depends on two factors:

(i)   the proportion of flux reaching the rotor from the lagging side of the pole compared to the leading side, and

(ii)   the point of flux division in the 'uniformly' magnetised annulus that determines the angle between the resultant, effective rotor mmf and the applied mmf.

Relative flux distribution in the pole sides

The B_r waveforms in the airgap which are a measure of the total 'useful' flux supplied by each pole give a clear indication of the concentration of flux on the two sides of the pole. Since the salient-pole construction of the field system is the main cause of maximum flux concentration at the pole tips, the effect of hysteresis and its intensity at a given excitation is exhibited by large peaks occurring in the waveforms at lagging pole tip B and relatively small peaks at the leading pole tip A; see for example Fig.C3.5, Chapter 3. Qualitatively, the greater the difference in the heights of the two peaks, the more pronounced the hysteresis effect; or it is the 'difference' part in the B_r waveform under the pole that is responsible for production of torque as depicted in Fig.C4.24. The remaining portion which is symmetrical about the pole axis would contribute little to the developed torque. Accordingly, the difference Δh in the peak heights would appear to be directly related to the torque.

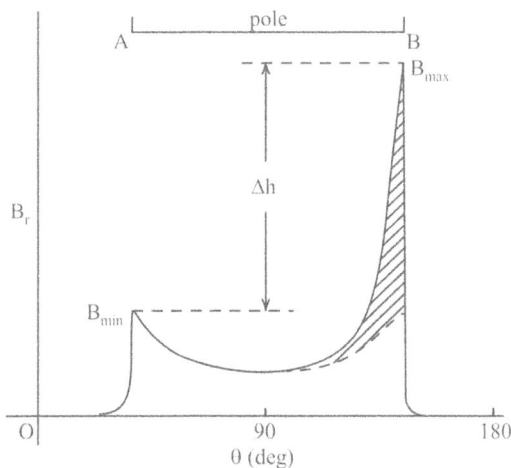

FigC.4.24 : B_r waveform and difference of peaks at two pole tips

A plot of Δh and developed torque vs. excitations is given in Fig.C4.25 showing a remarkable similarity between the two.

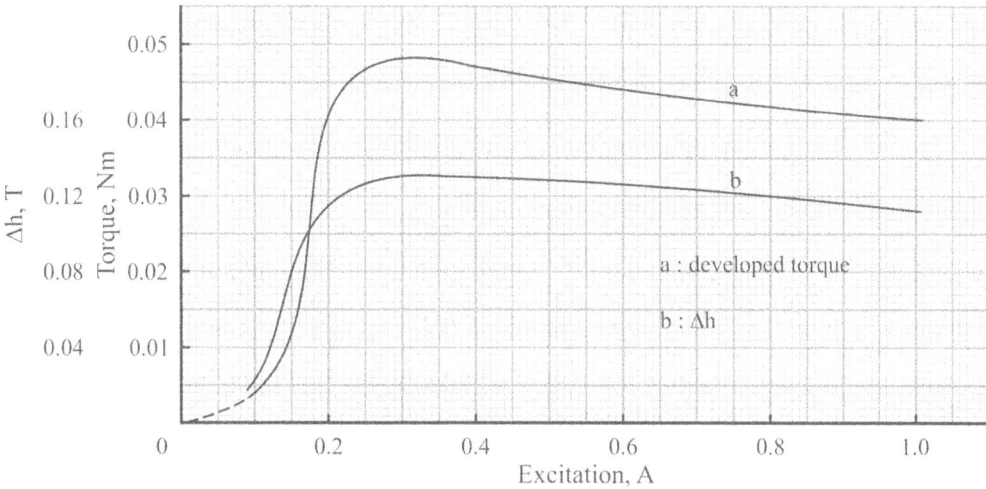

Fig.C4.25 : Δh and developed torque vs. excitation

The maximum value of Δh is seen to occur at an excitation of about 0.35 A and is comparable to 0.3 A being the excitation for maximum developed torque; at excitations beyond this, the trend in reduction of torque is also similar. A direct, linear dependence of developed torque on Δh is also reflected in the graph of Fig.C4.26.

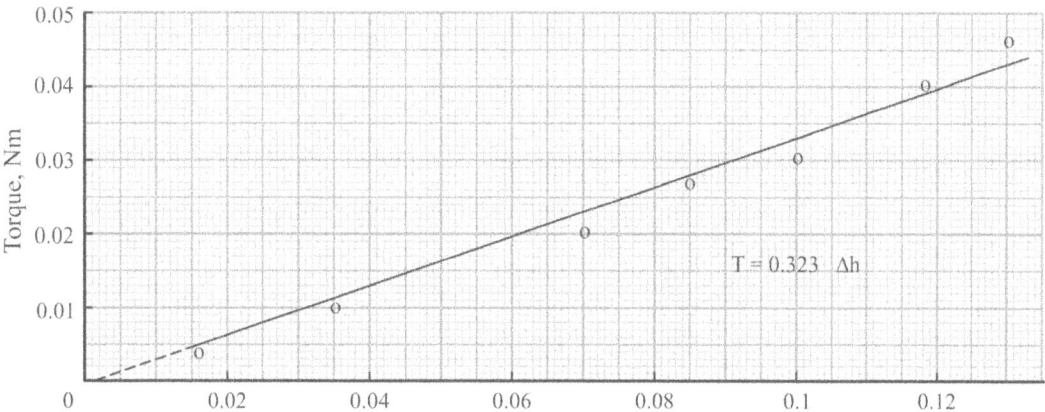

Fig.C4.26 : Variation of developed torque with Δh

Considering that a reduction in Δh is equivalent to a greater proportion of flux leaving the pole tip A compared with B, Fig.C4.24, the variation of Fig.C4.25 points to a "shift" in flux distribution in the pole. This concept originated in the first series of tests, using iron filings, to study the leakage flux distribution in the interpolar region[10]. That the reduction in torque cannot be due to a reduction in the total flux leaving the pole is demonstrated by the graph of Fig.C4.27 which shows a plot of flux/pole, represented by the area of the B_r waveform (for a unit axial length and fixed pole arc) against excitations. The lagging pole tip is heavily saturated at higher excitations and the shift of the flux away from it is to be expected.

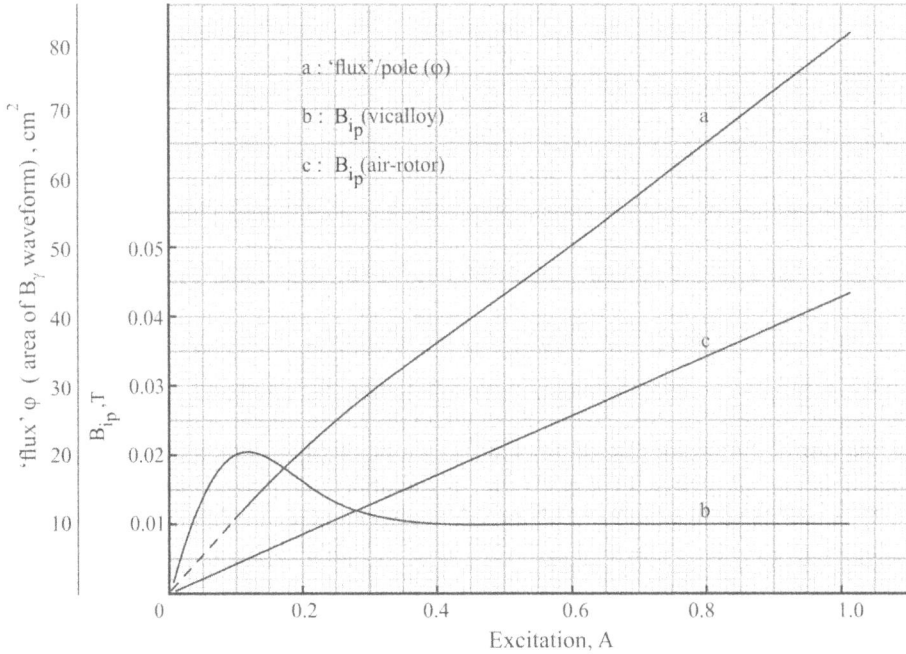

Fig.C4.27 : Flux/pole vs. excitation

As shown by graph b in Fig.C4.27, the flux density on the external rotor surface on the interpolar axis, B_{i_p}, exhibits a peak value thereby suggesting alternative paths for the flux. Below 0.3 A excitation, this variation of B_{i_p} (vicalloy), and also a slight curvature in the plot of flux/pole, ϕ, are indicative of relatively more flux entering the rotor corresponding to relatively higher permeability of vicalloy. The flux paths in the interpolar region tend to be more 'conventional' in this range of current, resulting in less leakage towards the yoke and higher flux density on the external rotor surface. Graph c shows the variation of B_{i_p} at the same point on the interpolar axis with the air rotor.

Change in direction of the rotor mmf

The significance of the point of flux division in the 'uniformly' magnetised annulus that determines the angle between the resultant, effective rotor mmf and the applied mmf

---

[10]See Appendix IV.

derives directly from the mechanism of torque production in terms of rotor magnetisation, $\overline{M}$. The point of flux division within the annulus and hence the direction of resultant rotor mmf shifts to the lagging pole tip in space with respect to the applied mmf even at low excitation and remains fixed at that position. Assuming a direct proportionality of $\overline{M}$ (B_θ under the pole surface) to developed torque, the torque-excitation curve should be similar to a plot of $\overline{M}$ with excitation; in other words to the graph of B_p against I, Fig.C4.4, which points to a constant torque at excitations greater than 0.3 A. Hence the reduction of torque should be equivalent to either a reduced value of $\overline{M}$ which is not possible or a shift of the rotor mmf, M_R, in the direction of the applied mmf, M_A. The second possibility would mean an increased torque in a 'negative' sense with a consequent reduction in the net torque as brought out in Fig.C4.28.

In this context, the role of the angle δ which MR makes with MA is that of a 'load angle'[11], the value of this angle affecting the torque observed at the shaft. Any reduction in this angle being tantamount to a reduction in the developed torque[12].

Clearly, the shift of flux from the lagging to the leading side of the pole that occurs at excitation currents in excess of 0.4 A can be interpreted in terms of diminishing of the "rotor hysteresis effect" leading to reduction of developed torque,

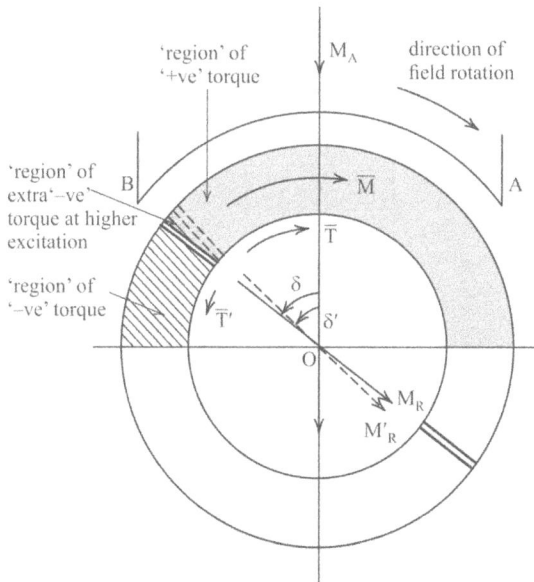

Fig.C4.28 : Regions of positive and negative torques

---

[11]Or 'torque angle' in a conventional sense.

[12]Although δ is almost constant in the experimental machine owing to the pole tip, a close examination of Fig.C4.3 reveals that the point of flux density reversal moves to the left, although only slightly, at increasing currents, that is from the lagging to the leading side of the pole.

Effect of rotational hysteresis

A phenomenon that is at times associated with the developed torque in hysteresis machines is the contribution from "rotational" hysteresis. This loss increases rapidly at low values of flux density, theoretically being twice as much as the alternating hysteresis loss[56], reaches a maximum near the knee of the saturation curve and then decreases to zero at intense saturation, a typical variation being as shown in Fig.C4.29 together with the alternating loss.

Under the assumption that both losses may contribute to the developed torque, the loss due to rotational hysteresis resulting from *directional* changes[43,44], it is not inappropriate to assume that the rotational loss is a consequence of radial flux density in the rotor annulus. However, it is more realistic to contemplate the effect as a result of directional changes in the 'small' elements that comprise the annulus, on the basis of the single particle theory of magnetism[45]. This approach is consistent with that explaining the production of torque by replacing the rotor magnetisation by magnetic dipoles.

In the experimental machine, changes in the directions of any particle of the rotor occur principally under the pole *tips,* taking it through one full cycle for every complete revolution of the rotor. This is evident from the B_θ waveform for vicalloy shown in Fig.C4.3. Under the leading pole tip the changes will be merely oscillatory about a mean position, the angle of swing not reaching even $90°$. Under the lagging pole tip the flux density in the vicalloy reverses and it is here that the particle rotates through $180°$. Rotation through another $180°$ takes place opposite the other lagging pole tip to complete the cycle.

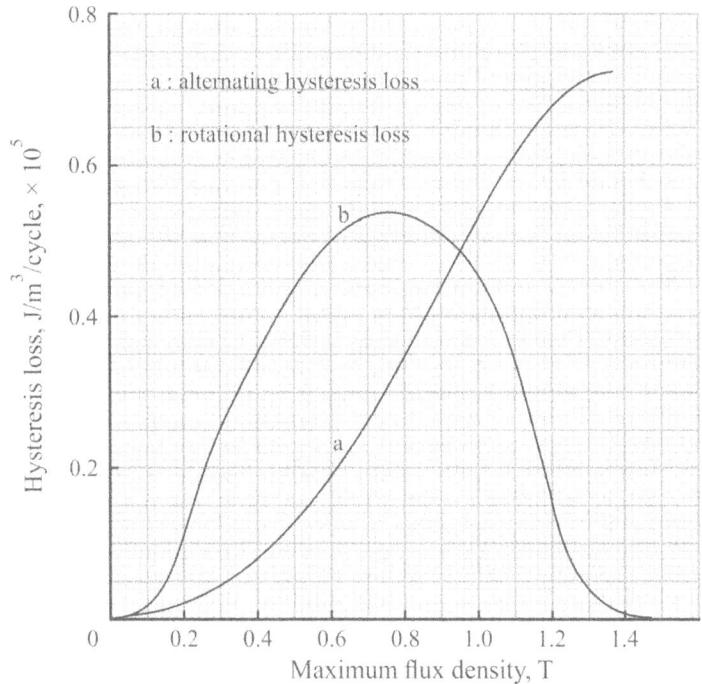

Fig.C4.29 : Rotational and alternating hysteresis loss

Although these conditions may give rise to rotational hysteresis loss, the flux density 'level' of the particle is in the intense saturation state for excitation currents more than even 0.15 A, reflected by appreciable peaks in B_θ waveforms, and therefore the rotational hysteresis loss and corresponding contribution to torque would be insignificant. It is observed that the maximum developed torque in the experimental machine is very nearly equal to the theoretically calculated value due to pure alternating loss[13]. Clearly, the rotational loss does not contribute to the developed torque in the experimental machine in the 'working' range of excitation and, accordingly, not contributory to the reduction of torque in region III. This is also borne out from the use of a thin rotor annulus where the magnetisation of the latter is predominantly peripheral with negligible radial component of flux density within[14].

If it is assumed that only alternating hysteresis would occur, as suggested by the use of a very thin rotor annulus, then the predicted values of torque would seem to be pessimistic. Thus, inevitably some rotational hysteresis loss might occur because of the rotating plane of the magnetic field (system), principally in the vicinity of the lagging pole-tips. This follows since the upper section of the torque curve in higher excitation range is quite different from the alternating hysteresis loss curve: the loss density increases with increasing excitation whereas the torque decreases. Clearly, in the initial part of the section, some contribution might occur due to rotational hysteresis in a small volume of the annulus[15].

Magnetic Scalar Potential Distribution in the Airgap

Computed waveforms

The significance of rotor magnetisation on the machine performance is also reflected in the form of magnetic *scalar* potential distribution on the external surface of the rotor, in the airgap. It is best derived from the measured *radial* flux density waveforms[16] and the computed waveforms are shown in Figs.C4.30 and C4.31.

The variations in Fig.C4.30 corresponds to an excitation of 0.3 A for both vicalloy rotor and air rotor. The waveform for the air rotor resembles that for the vicalloy rotor under non-hysteretic condition since the relative permeabilities in the two cases are much lower compared to very high permeability of poles and yoke.

Note that the simple, almost trapezoidal distribution of potential without spatial hysteresis, represented by the waveform for air rotor, is modified significantly to a characteristic triangular shape, with slight curvature in the interpolar region, by the *rotor*

---

[13]See Appendix I.

[14]In hysteresis machines designed with appreciably 'thick' rotor annuli, significant component of radial flux density may be present with the possibility of a good proportion of developed torque being due to rotational hysteresis loss with a consequent reduction of torque at high excitation under saturation.

[15]See, for example,

G.Wakui: Alternating hysteresis and rotational hysteresis in the hysteresis motor, Electrical Engineering in Japan, Vol. 90, (4), 1970, pp 85-93.

[16]See Appendix V for analytical treatment.

magnetisation. The striking feature of the waveform with the vicalloy rotor is the inclined straight-line portion stretching from the leading to the lagging pole tip[17]. This feature is common for all excitations, with the slop of the straight portion becoming steeper at increasing excitation currents.

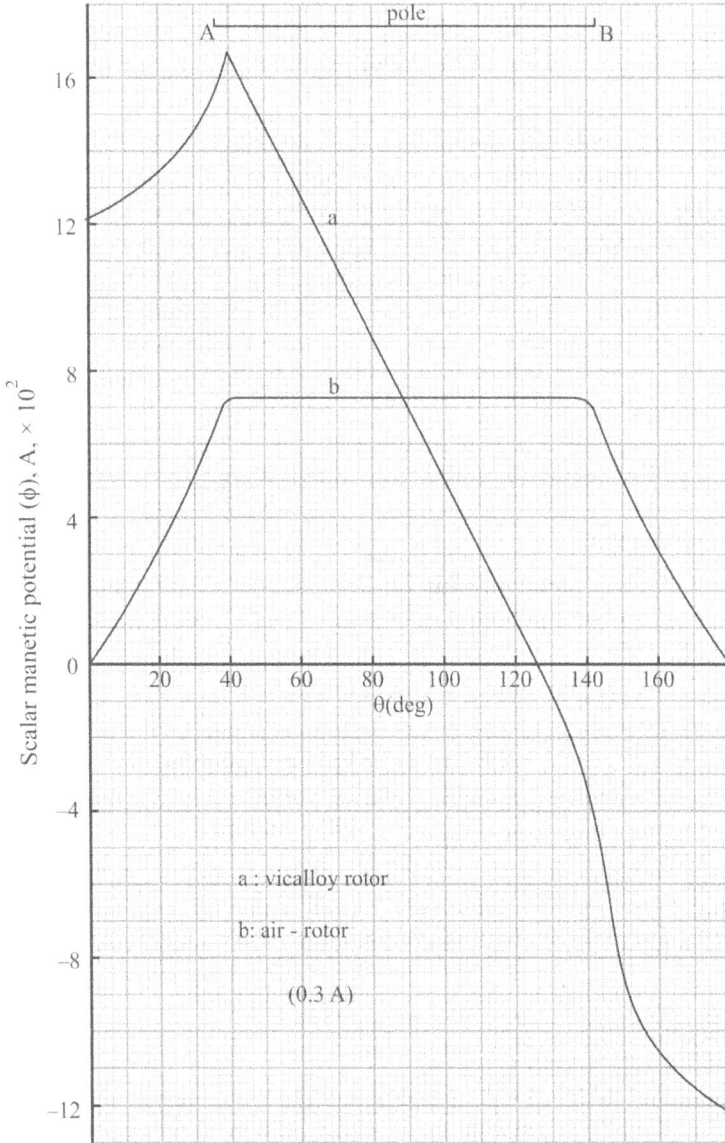

Fig.C4.30 : Magnetic scalar potential distribution at 0.3 A excitation

---

[17]Mathematically, this drives directly from the B_θ waveforms on the *external* rotor surface which have a constant, non-zero value under the poles, Fig.C3.5, Chapter 3.

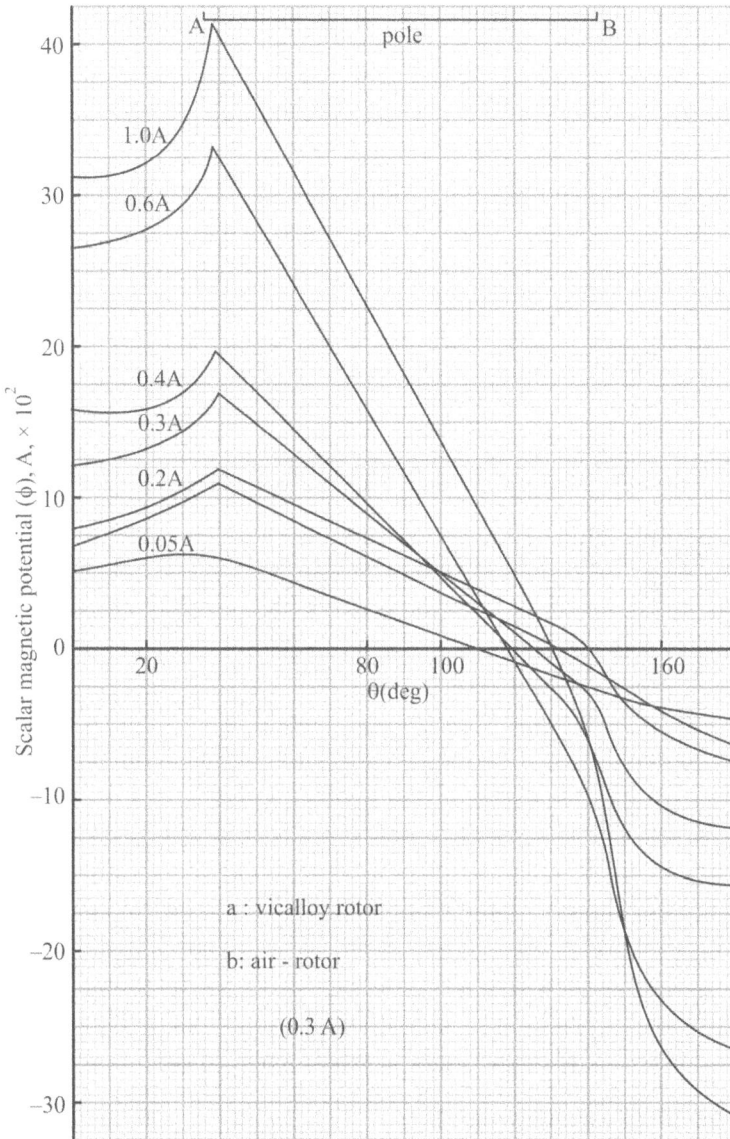

Fig.C4.31 : Magnetic potential distributions at excitations in the range 0.05 A – 1.0 A

Potential distribution in the airgap 'due to rotor magnetisation'

If the waveforms of the potential distribution with the air rotor are assumed to be representative of those obtained from the vicalloy rotor alone under non-hysteretic condition, then the effect of rotor magnetisation can be perceived by studying the waveforms shown in Fig.C4.32. These show the difference of potential between the air and vicalloy rotors for the same radius and excitation currents.

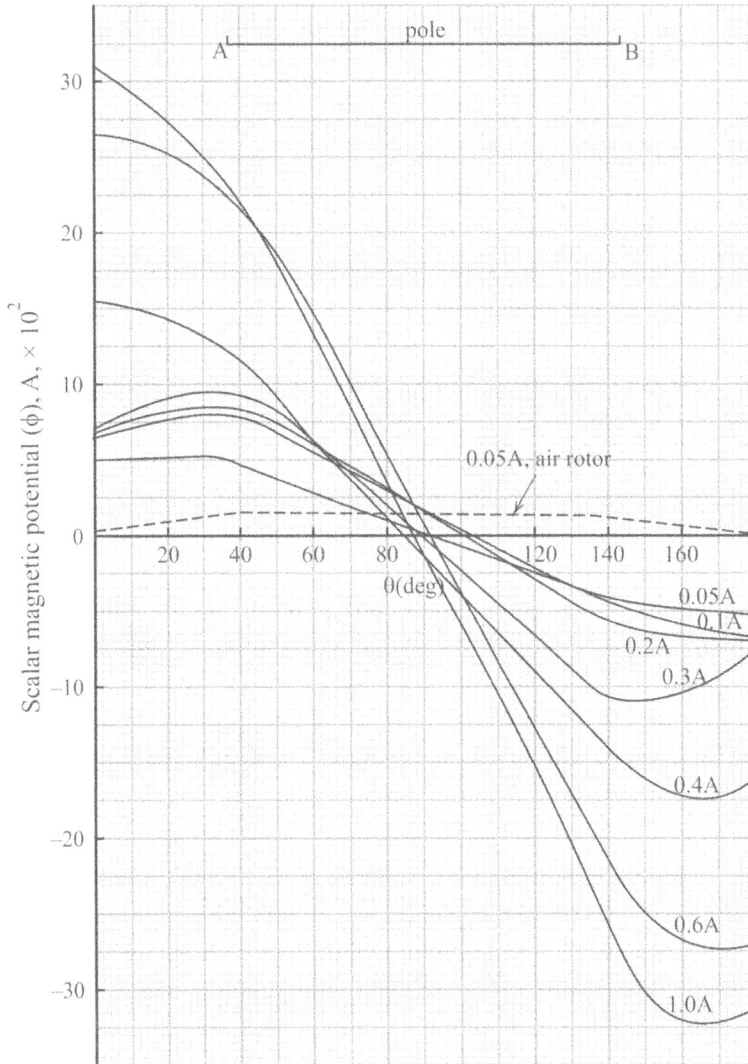

Fig.C4.32 : Magnetic scalar potential for air and vicalloy rotors

The two features of the waveforms are

(i) at a very low excitation current, 0.05 A, when the spatial hysteresis due to vicalloy is too feeble, the shape of the waveform approaches a trapezoid, but is displaced in space by approximately $90^{\circ}(E)$ from the one for the air rotor. This emphasises the 'independent' nature of rotor magnetism or magnetisation once magnetised[18].

---

[18]Experimental evidence of this effect is revealed in the iron filing patterns described in Appendix IV, exhibiting the residual magnetism of the rotor immediately after switching off the excitation.

(ii) for increasing excitation currents, the rotor potential is gradually modified in shape in the interpolar region, approaching a near-triangular variation, similar in shape to the armature reaction mmf waveform in an ordinary DC machine. The waveforms cross zero mark on abscissa at approximately 90° position for all currents > 0.2 A, supporting the arguments extended in (i) above[19].

Thus, rotor magnetisation is the key phenomenon to explain physical operation of the machine and production of torque. In terms of the fundamental component of the rotating mmf, the theory expounded in the foregoing should apply to operation of a hysteresis motor, too. The theory also is unaffected by the limitations arising from 'short' axial length of the machine, excluded from discussions at various stages. The 'end effects' caused by the short length may, however, change the field distribution slightly by introducing variations in the axial direction.

Iron Filing Patterns[20] and Developed Torque

Iron filing patterns obtained in the interpolar region of the machine at varying excitations bring out the distorting effect of *rotor magnetisation*. At low values of excitation it is hardly noticeable; above 0.15 A it becomes more apparent, reaching a maximum at the excitation for maximum developed torque and then decreasing sharply so that above 0.5 A it is hardly noticeable. When repeated for 'thick' rotor annuli, the excitation at which maximum distortion would occur is independent of rotor thickness. Also, similar patterns obtained in the *arbor* region show almost no variation with excitation.

The patterns suggest that the 'fall-off' in the torque is linked with changes in the rotor magnetisation near to the external surface of the rotor ring – a conclusion that can be drawn from the ΔB_r waveforms in the airgap. [See Figs.C2.5. 2.6 and C2.8, Chapter 2].

'Surface' Poles

In terms of a concept of fictitious 'surface' poles formed on the external rotor surface as a result of spatial hysteresis, the intensity of the surface poles opposite the lagging pole-tips increases with increasing excitation; whereas the intensity of 'other' surface poles initially decreases, it is imperceptible from 0.15 A to 0.25 A, and then increases. At excitation of 1.0 A, the relative amplitudes of the peaks on the waveforms are comparable with those at 0.1 A. This is owing to the different saturation magnetisations for the rotor (about 1.43 T) and field poles (2.1 T approx.). The intensity of the surface poles affects the peak values of the airgap B_r waveforms and, remarkably, the difference in the magnitudes is very nearly proportional to the developed torque. Clearly, the surface poles opposite the leading pole-tips are not much effective in any way over the normal operating range of the machine.

---

[19]As first approximation, the slope of the almost triangular mmf waveform due to rotor magnetisation can be related to the angle of hysteretic advance corresponding to a given excitation.
[20]See Appendix IV.

5 : Torque Calculation using Poynting Theorem

5

Torque Calcuation using Poynting Theorem

An exact, 'accurate' calculation of developed torque in a hysteresis machine would necessitate a precise knowledge of B and H distribution *within* the rotor annulus which may not be possible. However, the magnetic conditions of the rotor are distinctly reflected in the adjacent non-magnetic regions and this provides an alternative approach for computing power flow into the rotor and hence the torque at the shaft. This approach uses application of Poynting theorem when it is possible to define an appropriate Poynting vector and 'surface' for the integration of the vector such that power flow on the external and internal surface can be computed and hence the net power flow[1].

The Poynting Theorem

According to Poynting theorem[46], the vector

$$\bar{S} = \bar{E} \times \bar{H} \qquad (C5.1)$$

defines the instantaneous rate of flow of energy through unit surface area[2]. Since the units of $\bar{E}$ and $\bar{H}$ are V/m and A/m, respectively, the unit of S is W/m^2 and total power flow across a given or defined surface is obtained by integrating the *normal* component of S over the appropriate closed surface. That is

$$\text{total power} = \oiint \bar{S} \cdot \bar{i}_n \, ds = \oiint (\bar{E} \times \bar{H}) \cdot \bar{i}_n \, ds \qquad (C5.2)$$

where ds is an elemental area of the surface on which S is defined and $\bar{i}_n$ is the unit normal.

When evaluating power flow using the theorem, it is necessary to consider that

(a) according to eqn.(C5.2), it is only the total surface integral of $\bar{S}$ which gives the net power flow across a closed surface although ordinarily a 'useful' interpretation of

---

[1]See, for example,

S.C.Bhargava: Energy flow in a 2-pole hysteresis coupling by Poynting theorem, IEE Proc., Vol.130, Pt. A, No. 6, Sept 1983, pp 301-305.

[2]See Appendix VI for a derivation of the vector and its alternative forms.

$\overline{S}$ (= $\overline{E} \times \overline{H}$) be assumed to be power density at a point on the surface

(b) the postulate (a) implies that $\overline{S}$ is not a uniquely defined vector; any other vector whose integral over the closed surface vanishes can be added to $\overline{S}$ without altering the total power given by eqn.(C5.2)[3].

Poynting Theorem Applied to the Experimental Machine

Regions of power flow

A reference to the constructional features and mode of operation of the experimental machine reveals that the 'power' crossing the airgap does NOT evolve essentially from the field system as in a conventional hysteresis motor since the former is driven mechanically and excited separately by direct current. It is the combination of the two facets (mechanical rotation of DC excited field system), with the DC excitation playing a more crucial role, that causes hysteresis loss in the annulus and production of torque[4].

To illustrate the above concepts further, a simple 'block' diagram of the machine is given in Fig.C5.1, showing the main regions of energy flow.

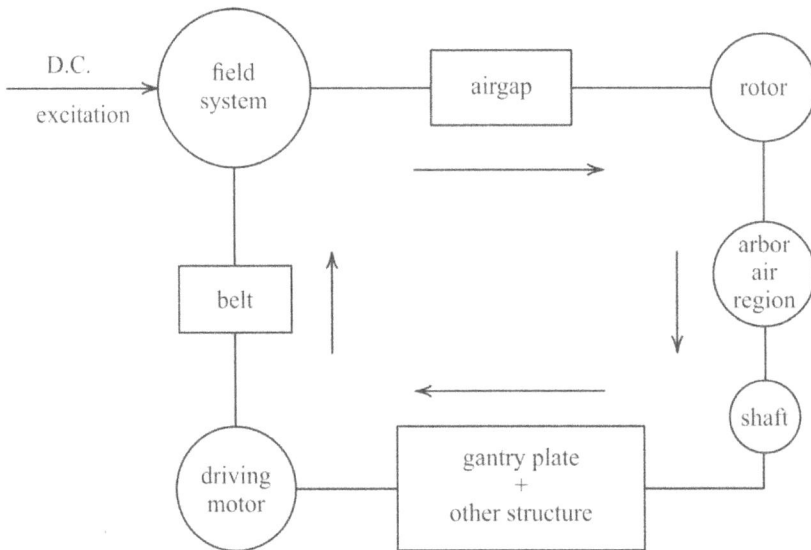

Fig.C5.1 : Block diagram of the experimental machine

[3]A common example is curl of a vector; in some cases this permits a more convincing interpretation of the power crossing the closed surface[47]. However, there is little to be gained by such an addition when only the total power is of interest.

[4]In this context, the machine can be identified more easily as a hysteresis *coupling* than a motor as mentioned before.

The driving motor supplies mechanical power to the rotating field system through the belt. In a simple device this power would simply equal the mechanical losses. However, when electric power is supplied to the field system as well, energy crosses the airgap towards the rotor and 'returns' to the (driving) motor via the arbor region, the rotor shaft, a the gentry plate and other mechanical structure of the assembly. Some of this energy is used to supply the rotor hysteresis loss[5].

If a Poynting vector defined by eqn.(C5.1) is adopted, with the measurement of $\bar{E}$ and $\bar{H}$ carried out appropriately, then the integral over a closed surface enclosing the rotor would provide the net power entering the annulus that would represent the alternating hysteresis loss in the present case[6].

The Two Rotor Surfaces

From the aspect of measurement of the component vectors $\bar{E}$ and $\bar{H}$, the simplest method to obtain the net power flow into the rotor active material or the annulus is to integrate the Poynting vectors independently on all the surfaces of the annulus. With a proper sign convention, this would account for the net power flow entering the rotor from the airgap and leaving it through the arbor region in accordance with the diagram of

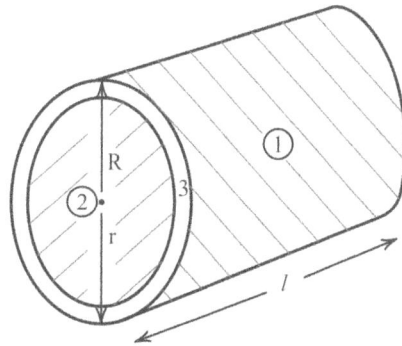

Fig.C5.2 : Surfaces of the rotor for integration

Fig.C5.1. As illustrated in Fig.C5.2 the total surface for integration will comprise the two cylindrical surfaces 1 and 2 and the annular discs 3 at each end. The first two being referred to as external and internal surfaces of radii R and r, respectively, the corresponding areas for an axial rotor length l will be

$$S_1 = 2\pi R l \quad \text{and} \quad S_2 = 2\pi r l \quad\quad (C5.3)$$

The areas of the two annular discs, 3, can be calculated similarly. However, since these will be extremely small owing to use of a very thin annulus when compared to either S_1 or S_2, the contribution of energy flow through them to the integral, eqn.(C5.2), can be ignored[7].

---

[5]Fig.C5.1 gives only a general sense of energy flow with disregard to various sinks and sources of power at different stages en route. However, referring to the air spaces adjacent to the rotor, the net power loss across them corresponds to the power dissipated in heat as hysteresis loss.

[6]This would imply a three-dimensional picture in general, on each element of which the vectors $\bar{E}$ and $\bar{H}$ are known. However, for fields invariant in the axial direction only 'external' and 'internal' rotor surfaces need be considered as in the present case.

[7]This is in line with the assumption that the field distribution in the experimental machine is essentially two-dimensional and any 'end effects' are disregarded.

Distribution of E and H on the rotor surfaces

As described, the basic quantities measured in the machine were the induced EMFs proportional to the radial components of flux density in the airgap and arbor regions. Thus, with a relative speed of rotation of the field system of ω rad/s, the induced EMF will be

$$e \; = \; B_r \, l \, \omega \, r \qquad\qquad (C5.4)$$

where B_r is the radial component of flux density, l the total axial length of the search coil (used for B_r measurement) and r the radius.

Eqn.(C5.4) allows the distribution of the electric field intensity to be deduced. Using cylindrical coordinate system, the only component of $\overline{E}$ applicable is in the axial direction for a constant peripheral velocity v ($= \omega$ r) at a given radius r. Hence

$$\overline{E} \; = \; E_z \, \overline{i_z} \qquad\qquad (C5.5)$$

where $\overline{i_z}$ denotes the unit vector in the axial or z-direction and E_z is obtained from eqn. (C5.4) as e/l.

The distribution of $\overline{H}$ on the corresponding surface is obtained from the knowledge of $\overline{B}$ at different points; since the measurements are made in non-magnetic regions – airgap or arbor – $\overline{H}$ can be deduced from $\overline{B} = \mu_o \, \overline{H}$. More specifically, the tangential component of $\overline{H}$ is required in association with $\overline{E}$ above to give power flow in the radial direction or across the annulus as depicted in Fig.C5.3.

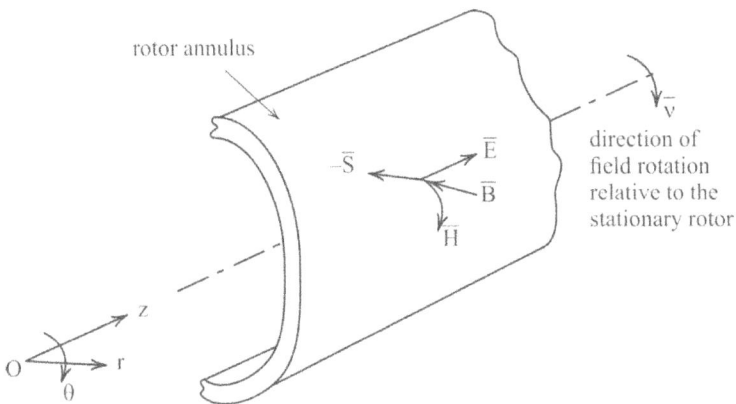

rotor annulus

direction of field rotation relative to the stationary rotor

Fig.C5.3 : Components of $\overline{B}$ and $\overline{H}$ and rotor surface

This component is derived from the B_θ variation obtained experimentally and given by

$$H_\theta = B_\theta / \mu_o \qquad\qquad (C5.6)$$

With these components known, the appropriate Poynting vector is given by

$$S_r \; \bar{i}_r = - E_z \, H_\theta \; \bar{i}_r \quad \text{(that is, into the rotor)}[8] \tag{C5.7}$$

Since in the experimental machine B_r is measurably non-zero only under the poles, being negligible in the interpolar region, a necessary condition for radial power flow is a non-zero 'average' value of H_θ under the poles, stressing the importance attached to B_θ waveforms in the preceding chapters.

Comment

Clearly, if B_r were symmetrical about the pole axis and H_θ symmetrical about the interpolar axis, no net power flow would be possible. This can be demonstrated as follows:

Let B_r be expandable in Fourier series of the form

$$B_r = k' \sum_{n=1,3...}^{\infty} B_{m_n} \sin n\theta \tag{C5.8}$$

giving E_z as

$$E_z = k \sum_{n=1,3...}^{\infty} B_{m_n} \sin n\theta \tag{C5.9}$$

and H_θ as

$$H_\theta = k'' \sum_{n=1,3...}^{\infty} H_{m_n} \cos n\theta \tag{C5.10}$$

Then the radial Poynting vector will be

$$S_r = -E_z \, H_\theta = -k \, k'' \sum_{n=1,3...}^{\infty} \sum_{n=1,3...}^{\infty} B_{m_n} H_{m_n} \sin n\theta \cos n\theta \tag{C5.11}$$

and its integrated value over a closed surface of radius r will be zero. This shows that no net power transfer can be obtained for the rotor under non-hysteretic conditions or with the air rotor.

Non-dependence of net power flow on radius

A particularly important property of S_r as defined in eqn.(C5.7) is that the calculation of the total power, that is the integral of S_r over the closed rotor surface, is independent of radius. This can be proved in terms of general expressions for B_r and B_θ derived from a magnetic *scalar* potential ϕ and speed of rotation of the field system, ω.

---

[8] The negative sign in $\overline{S}$ obtains because the direction of $\overline{S}$ should be outward to the surface according to the convention; the normal to the surface being always outward.

Consider the airgap region, Fig.C5.4, where ϕ is governed by Laplace's equation and its variation expressed as a series of harmonically varying terms.

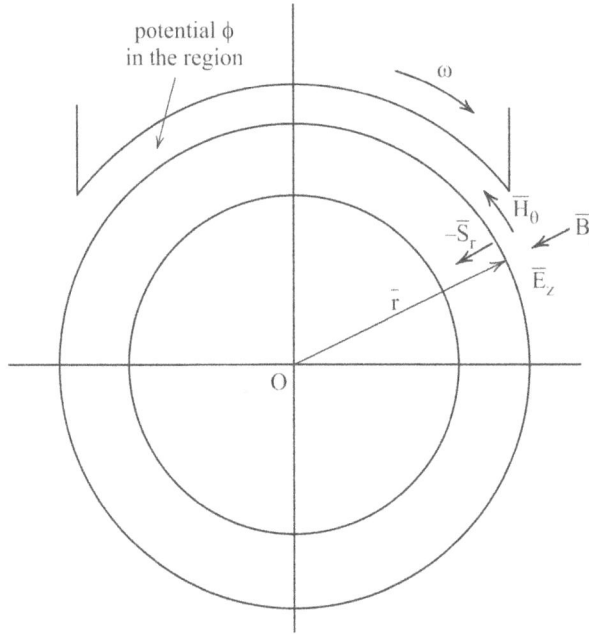

Fig.C5.4 : Magnetic scalar potential in the airgap region

Hence

$$\phi = \left(A_n\, r^n + B_n\, r^{-n} \right)\sin n\theta + \left(C_n\, r^n + D_n\, r^{-n} \right)\cos n\theta \qquad (C5.12)$$

The radial flux density B_r is given by

$$B_r = \mu_0 \mu_r = -\mu_0 \frac{\partial \phi}{\partial r} \qquad (\text{from } \overline{H} = -\overline{\nabla}\phi)$$

or

$$B_r = \mu_0 n \left[\left(A_n\, r^{n-1} - B_n\, r^{-n-1} \right)\sin n\theta + \left(C_n\, r^{n-1} - D_n\, r^{-n-1} \right)\cos n\theta \right] \qquad (C5.13)$$

and the induced EMF in the axial direction

$$e_z = B_r\, \omega\, l\, r \qquad \text{giving} \qquad E_z = B_r\, \omega\, r$$

or $\quad E_z = -\mu_0 n\, \omega\, r \left[\left(A_n\, r^{n-1} - B_n\, r^{-n-1} \right)\sin n\theta + \left(C_n\, r^{n-1} - D_n\, r^{-n-1} \right)\cos n\theta \right]$ (C5.14)

Similarly, H_θ is obtained from eqn.(C5.12) as

$$H_\theta = -\frac{1}{r}\frac{\partial \phi}{\partial \theta}$$

$$= -n\left[\left(A_n\, r^{n-1} + B_n\, r^{-n-1} \right)\cos n\theta - \left(C_n\, r^{n-1} + D_n\, r^{-n-1} \right)\sin n\theta \right] \qquad (C5.15)$$

From eqns.(C5.14) and (C5.15)

$$S_r = -E_z H_\theta$$

$$= -K_1 r \left\{ A_1 \sin^2 n\theta + A_2 \cos^2 n\theta + A_{12} \sin n\theta \cos n\theta \right\} \qquad (C5.16)$$

where

$$K_1 = \mu_0 n^2 \omega$$

$$A_1 = -\left[A_n C_n r^{2n-2} - B_n C_n r^{-2} + A_n D_n r^{-2} - B_n D_n^{-2n-2} \right]$$

$$A_2 = \left[A_n C_n r^{2n-2} + B_n C_n r^{-2} - A_n D_n r^{-2} - B_n D_n^{-2n-2} \right]$$

$$A_{12} = \left[A_n^2 r^{2n-2} - B_n^2 r^{2n-2} - C_n^2 r^{2n-2} + D_n^2 r^{-2n-2} \right]$$

When integrated over the cylindrical rotor surface, the last term on the right-hand side of eqn.(C5.16) will vanish and the total power given by

$$P = -K_1 K_2 r \int_0^{2\pi} \left(A_1 \sin^2 n\theta + A_2 \cos^2 n\theta \right) r \, d\theta$$

or $\qquad P = -K' r^2 (A_1 + A_2) \qquad (C5.17)$

where K_1, K_2 and K' are constants.

Substituting for A_1 and A_2 in eqn.(C5.17)

$$P = K [B_n C_n + A_n D_n], \quad K = \text{constant} \qquad (C5.18)$$

This shows that the total power is only a function of the coefficients A_n, . . . , D_n and NOT of radius r. This is to be expected since no net power is absorbed in the 'air' spaces. It is also clear that the expressions for E_z and H_θ must be representable in the form given by eqns.(C5.14) and (C5.15) in order to yield net power flow; in other words, the coefficients of sin nθ and cos nθ terms should be independent of each other and functions of r.[9]

Poynting vector in the peripheral direction

The existence of a component of $\overline{H}$ in the radial direction indicates a second Poynting vector in the peripheral direction on the rotor surface given by

$$S_\theta \, \overline{i_\theta} = E_z H_r \, \overline{i_\theta} \qquad (C5.19)$$

This has a mathematical, non-zero value on each elemental area of either surface and, being only circulatory flow of energy, does not contribute to the power flow in a radial direction or into the rotor annulus.

---

[9]The non-dependence of total power on the radius of calculations has a useful bearing on the principle behind the measurement of power flow using search coils in a particular region: so long as the radii of the two coils are accurately known, apparently to calculate B_r values, their exact position is not critical.

Computation of Power Flow

Using the measured waveform of E_z and *computed* waveform of B_θ, or H_θ, let these be expressed as

$$E_z = K_1' \sum \left\{ E_{1_n} \sin n\theta + E_{2_n} \cos n\theta \right\}$$

and
$$H_\theta = K_2' \sum \left\{ H_{1_n} \sin n\theta + H_{2_n} \cos n\theta \right\}$$

at any excitation and for n^{th} harmonic.

Then the radial Poynting vector for the computation of power flow would be
$$S_r = - E_z H_\theta$$

$$= - K_1' K_2' \sum \left\{ E_{1_n} \sin n\theta + E_{2_n} \cos n\theta \right\}$$

$$\times \sum \left\{ H_{1_n} \sin n\theta + H_{2_n} \cos n\theta \right\}$$

and the total power flow on either rotor surface (for which $E_{1_n}, \ldots, H_{2_n}$ apply) is

$$P = K'' \sum \left\{ E_{1_n} H_{1_n} + E_{2_n} H_{2_n} \right\}^{10} \tag{C5.20}$$

where K'' is a constant.

The net power absorbed in the rotor

The results of power flow computation using measured waveforms for both external and internal rotor surfaces are plotted against excitation in Fig.C5.5.

The power flow on either surface increases sharply with excitation, being almost proportional to currents beyond 0.6 A. This trend is understandable as mathematically the integrated Poynting vector is the summation of the product of B_r- and B_θ-series coefficients. As shown by the waveforms of Figs.C3.2 and C3.5, Chapter 3, the magnitudes of these flux density components under the poles increases steadily with increasing excitation. Although the radial component is small on the inner surface, the peripheral component is relatively large compared to that on the external surface and hence the product in the form of power flow given by graph II in Fig.C5.5 closely follows the variation on the external surface of the rotor.

The power flow *into* the rotor representing (alternating) hysteresis loss or the developed torque is the difference of the two computed power flows. Observe that the power represented by curve I enters the rotor from the airgap and leaves the rotor through the arbor, showing that only a small percentage of the total power entering the rotor is 'utilised', the rest returning to the shaft. See Fig.C5.1.

---

[10]See Appendix V for details of (computer) calculation and evaluation of constant K'' to give power flow in synchronous watt.

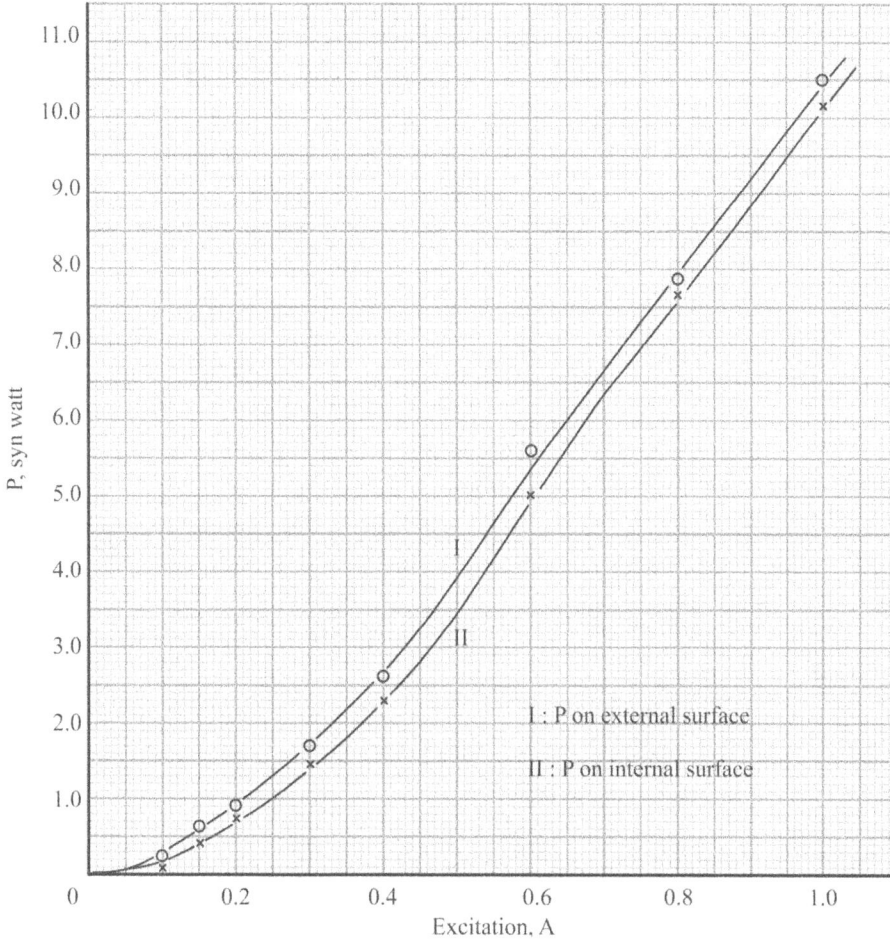

Fig.C5.5 : Computed power flow on external and internal surfaces of the rotor

'Shape' of the Poynting vector

The Poynting vector at a particular point on the rotor surface has no distinct meaning since it is only an integrand and any other vector having zero divergence over the whole surface can be added to it. Mathematically, therefore, any point-to-point interpretation of the vector evaluated during calculations is not truly valid. Nevertheless, a 'qualitative' picture of the power flow at different points on the surface can still be conceived from the plot of the vector around the rotor periphery.

Three illustrative plots of Poynting vector for the external rotor surface at excitations of 0.1 A, 0.3 A and 1.0 A are given in Fig.C5.6(a) and (b).

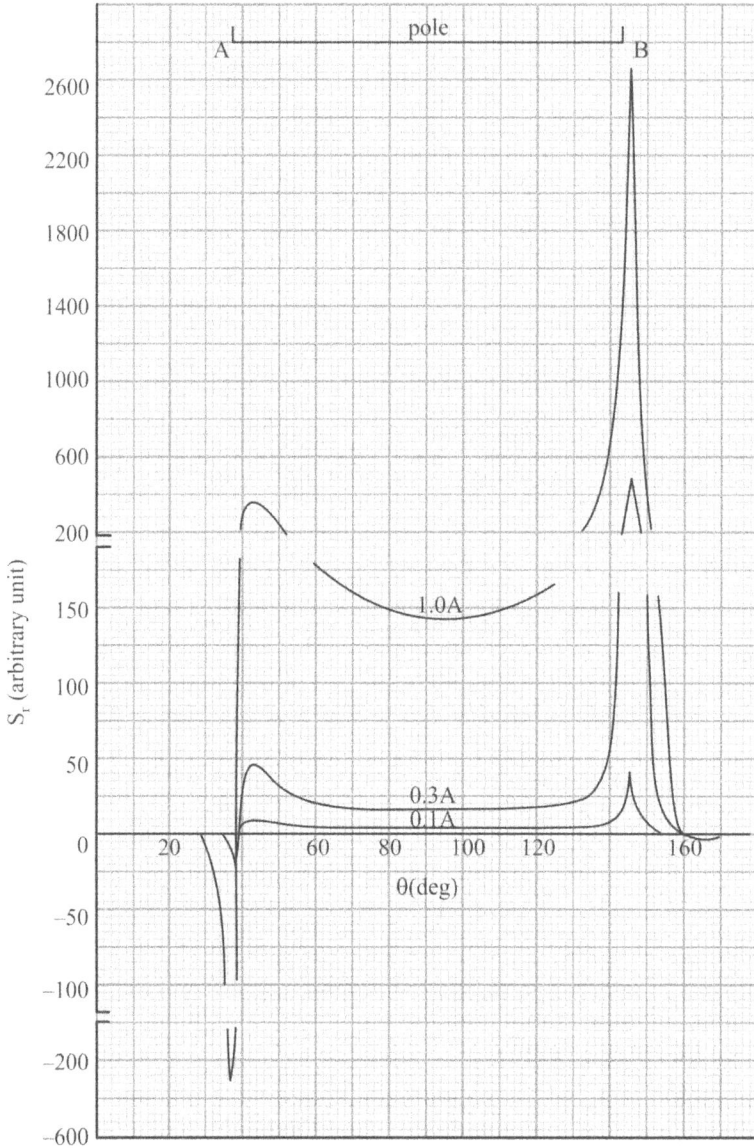

(a) external surface

Since the direction of both E_z and H_θ reverses under each pole, the plots in the figures apply over either pole pitch and the power flow under the pole arc is always positive as shown. In spite of the magnitude of the vector changing considerably at increasing excitation, the shape does not change much. Also, the magnitude at any excitation is greatest at the lagging pole tip confirming that a large proportion of power flow takes place from these pole tips as brought out in previous chapters, and matching the sharp peaks observed in the waveforms of B_r and H_θ at lagging pole tips.

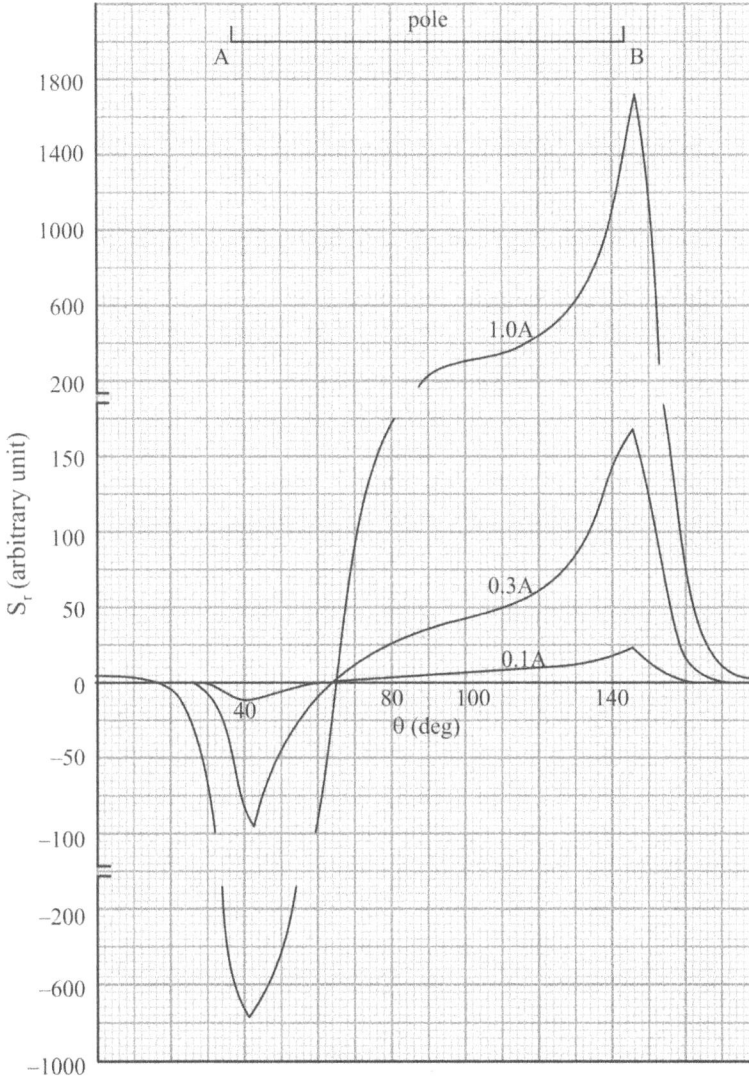

(b) internal surface

Fig.C5.6 : Shape of Poynting vector

The plots indicate *negative* values of power flow opposite leading pole tips and if the previous interpretation to hold, this means that energy leaves the external rotor surface to return to the field system. This reversal is also indicated for the internal surface at similar currents shown in Fig.C5.6(b) where the relative height of the peak in the negative direction is much larger. This phenomenon, too, supports the explanation for the reduction of torque at higher excitations due to modification of flux distribution.

Effect of mmf harmonics

Whilst the effect of space harmonics in practical motors due to currents in three-phase distributed winding leads to complications in their analytical treatment mainly owing to relative speeds of rotation of harmonics, in the experimental machine the effect of harmonics is realisable using simple principle of superposition. A typical example is the power flow on the external rotor surface, computed for an excitation of 0.3 A for a range of harmonics from 1 to 100. The extent to which these contribute to the total power is recorded in Table C5.1.

Table C5.1: Contribution of mmf harmonics to power flow

Order of harmonic	Range of the order	Contribution as percentage of the fundamental, %	Total net contribution in terms of the fundamentals, %
1		100.00	
3		0.92	
5		14.92	
7		9.57	
9		2.31	
11		9.42	
13		0.71	
15		4.2	
17		2.8	
19		0.51	
21		3.1	
23		−0.04	
	3 − 23		48.4
	25 − 29		2.5
	31 − 39		1.26
	41 − 49		0.35
	51 − 99		0.78
	101 − 199		0.018
	3 − 199		53.0

It is observed that

    (i) the harmonics contribute a little over 50% of the total power, a large proportion being due to harmonics in the range 3 − 21;

    (ii) the effect of all the harmonics that contribute appreciably is additive to that due to fundamental alone[11].

---

[11]The fact that the relative contribution of the 5th harmonic is about 15 times that of the 3rd harmonic, rather in the reverse order, is due primarily to the pole arc of 106° or almost two-third of the pole pitch and the characteristic shapes of the B_r and B_θ waveforms.

Calculation of Torque from the Net Power Flow

Transference of power in the machine

The schematic of the machine and driving motor are shown in Fig.C5.7. The power from the motor, P_B, is transmitted through the belt and drives the field system at N rps. Applying Poynting theorem to the stationary annulus surfaces S' and S'' surrounding the field system and the rotor, respectively, a net energy flows into S' and emerges from S''. Hence if P_m is the net mechanical energy flow from the motor through S' and P_e through S'' then, neglecting any friction and windage losses,

$$P_e = P_m \qquad\qquad (C5.21)$$

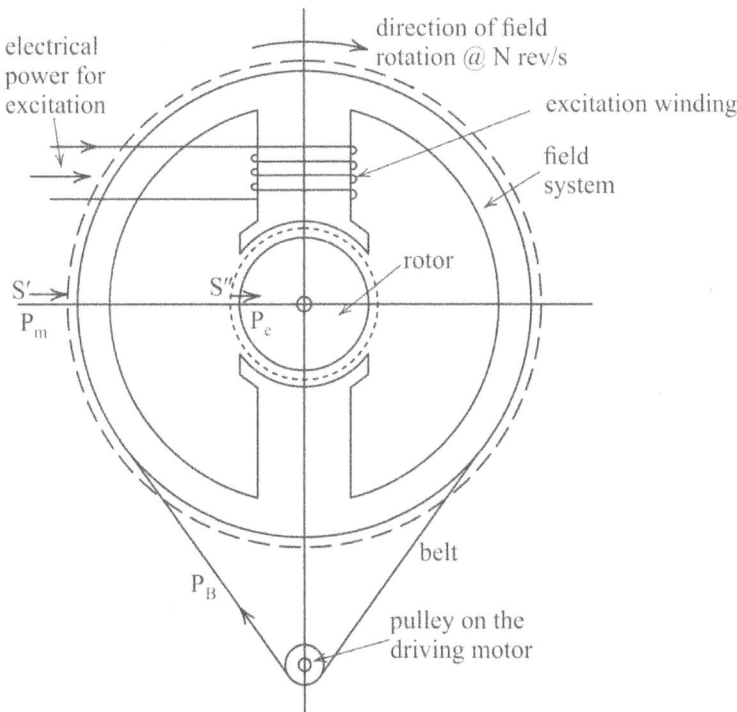

Fig.C5.7 : Schematic showing transference of power

The electrical power through S' is not included as this only supplies the copper loss of the excitation winding.

Since the rotor and the two surfaces S' and S'' are stationary with respect to the same reference frame, P_e must equal the power dissipated as heat in the rotor hysteresis loss.

Let the hysteresis loss per revolution be W_h W. Then the total loss/s will be N W_h W and

$$P_e = N W_h \qquad\qquad (C5.22)$$

The hysteresis loss, or P_e, is equal to P_m and manifests as torque T, exerted on the rotating field system because of the relative motion.

Hence, if the angular speed of rotation is ω rad/s,

$$P_m = \omega T$$

so that $\qquad\qquad T = P_m/\omega = P_e/\omega = NW/\omega \qquad\qquad (C5.23)$

or putting $\qquad\qquad N = \omega/2\pi$

$$T = W_h/2\pi \text{ Nm} \quad \text{ or } W_h \text{ W(syn)} \qquad\qquad (C5.24)$$

Eqn.(C5.24) shows that the developed torque is numerically equal to the hysteresis loss in the rotor ('active') material and is *independent of speed,* provided that the standstill condition is excluded.

Calculated and measured torque-excitation curves

The torque-excitation curve derived from the computed power, together with the measured curve, are shown in Fig.C5.8. Whilst the calculated torque values are in excess of the measured values, esp. at higher excitations, the shape of the two curves is similar; the correlation being satisfactory in the linear or normal working range.

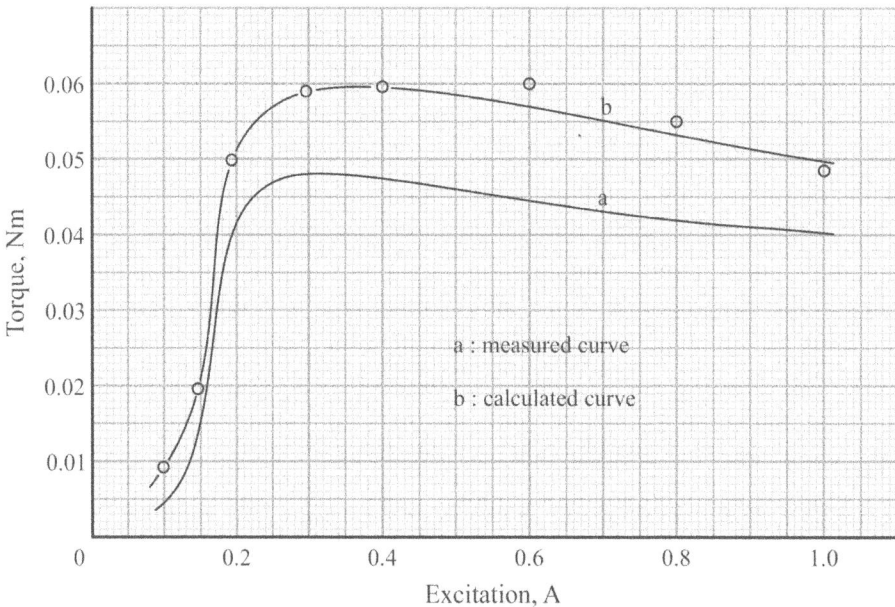

Fig.C5.8 : Computed and measured torque at various excitations

The error between the two curves is rather apparent for the calculated torque values are derived from the *difference* of two comparable values of power on the external and internal rotor surfaces. Notwithstanding, the three regions of the torque curve are in evidence, reflecting the overall accuracy of the Poynting theorem approach used for computation[12].

---

[12]In general, the discrepancy between the curves can be ascribed to an extent to experimental errors in the measured values of ΔB_r and H_θ themselves, esp. considering mechanical features of the full-pitch search coils, including the finite diameter of the wire used for winding the coils vis-à-vis the sharpness of coil-side contours.

See, for example,

S.C.Bhargava: Electrical Measuring Instruments and Measurements (book), Ch. XII, B S Publications, India, 2013.

APPENDICES

1

Experimental determination of hysteresis loops

Use of Hysteresigraphs and Permagraphs[1]

The hysteresigraph

Commencing with the need to obtain the 'virgin' B-H curve, the experiments on a sample of permanent magnet material are characterised by plotting of major and minor hysteresis loops and, at times, the various recoil loops. The testing methods may vary depending on the size and shape of the magnet and its application[2]

The most convenient and versatile device or means to obtain all test results as above, and meet all requirements as discussed previously, is a hysteresigraph, a variety of which are now commercially available. A hysteresigraph usable for various test is detailed schematically in Fig AP1.1[3].

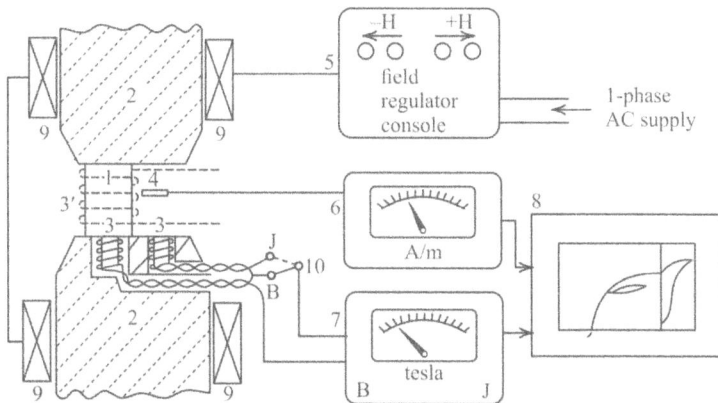

Fig.AP1.1 : Schematic of a hysteresigraph and its details

---

[1]See, in particular,

S.C.Bhargava: Electrical Measuring Instruments and Measurements (book), B S Publications (also CRC Press), Hyderabad, 2013, Chapter 12.

[2]For example, a small rectangular ALNICO magnet used as a brake magnet in a single-phase or three-phase induction-type energy meter.

[3]Developed and manufactured by Dr. Steingroever, GMBH, (of) Magnetic-Physic, Germany.

Description

1	:	sample under test
2	:	poles of the electromagnet
3	:	pole "B" coils
3′	:	"B" coil around the sample
4	:	Hall probe for measuring magnetising force
5	:	Regulator for controlling magnetising force in "+ve" and "-ve" direction
6	:	Indicator for magnetising force value, A/m
7	:	Meter to read flux density 'inside' the sample, for example a gauss meter
8	:	X-Y plotter or recorder for plotting B-H curve and loop(s)
9	:	Magnetising winding
10	:	Switch to adjust pole coil connection

A photograph of the hysteresigraph is given in Fig. AP1.2.

Fig.AP1.2 : The hysteresigraph

Process of measurement

The test piece or the sample[4], having been completely demagnetised, is positioned between the parallel poles of the electromagnet as shown in Fig.AP1.1. The position of the moveable upper pole is adjusted to suit the vertical height of the sample so that both the poles make good contact with the sample surfaces. The upper, movable pole is clamped in this position.

---

[4]Usually flat and round in shape, about 20 to 25 mm in diameter and 10 mm in 'height'; the flat surfaces having been machined very carefully and polished so as to make firm, good (magnetic) contact with the pole surfaces..

Measurement of Magnetising Force and Flux Density

The magnetising force *inside* the sample is sensed and measured with the help of a flat, transverse-type calibrated Hall probe, located as close to the sample as possible (see Fig.AP1.1). The Hall probe output is measured on a meter calibrated to indicate magnetising force in A/m.

The flux density in the sample corresponding to a given magnetising field can be measured in two different ways:

(a) by using a closely fitting search coil of suitable number of turns, N, wound round the sample ($3'$ in Fig.AP1.1)), the flux density being given by $B = \emptyset/A$ where $\emptyset$ is the flux produced and linking the coil and A its area of cross section, the flux being measured on a calibrated flux meter or otherwise;

(b) by means of "pole coils" embedded in the lower, removable pole piece of a diameter or cross-sectional area smaller than that of the sample, marked 3, 3 in Fig.AP1.1. The pole piece is provided with *two* identical coils, having same number of turns and cross-sectional area, connected to a flux meter via a special switch (marked 10 in Fig.AP1.1) so as to allow either one or both the coils to be connected in the circuit for the purpose considered later.

Basis of measurement of magnetising field and flux density using "B" coils

The provision of pole coils in the apparatus does away with the necessity of winding a search coil around the sample every time a new sample is to be tested.

Assuming that the pole pieces are made of very high permeability material[5] so that they do not get saturated even at the highest value of magnetised field used for measurement, the pole surface would be a (magnetic) equipotential and hence the normal component of *flux density* would be continuous across the common surface as shown in Fig.AP1.3(a). It follows that tangential component of *magnetising field* would

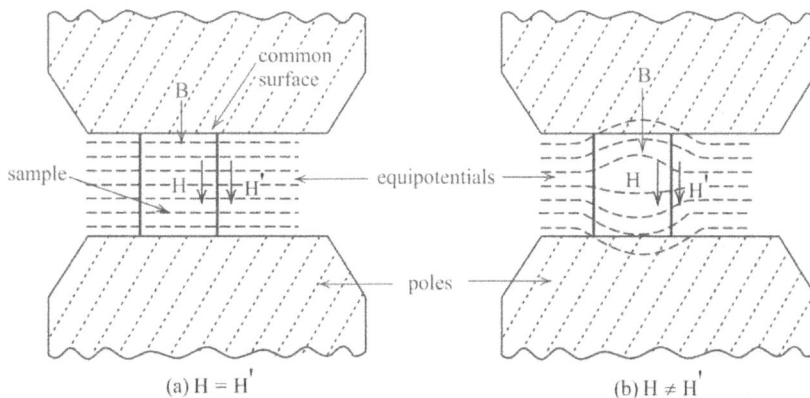

Fig.AP1.3 : Measurement of magnetising field and flux density using Hall probe and "B" coil

---

[5]For example, permendur.

likewise be continuous along the air-iron boundary by the side of the sample, thus justifying the use of the Hall probe to measure magnetising field *inside* the sample.

Operation to Plot a Hysteresis Loop

Once the sample is positioned between the poles, fully covering one of the coils (usually the left coil), the field regulator is adjusted to slowly increase the magnetising field in the "+ve" direction resulting in plotting of B-H curve of the material. At the limiting value of H, given by $+H_m$, the current is reversed and decreased gradually to reach the residual flux density position of the loop where H=0, and then reduced further to coercivity position, B=0. The magnetising field is continued to 'increase' in the negative direction till the point $-H_m$. $-B_m$ is reached as shown in Fig.AP1.4 to illustrate a qualitative hysteresis loop. Following this, the current is again reversed and increased in positive direction to close the hysteresis loop at $+H_m$, $+B_m$[6], although this part of the loop may not, in general, be required to be plotted.

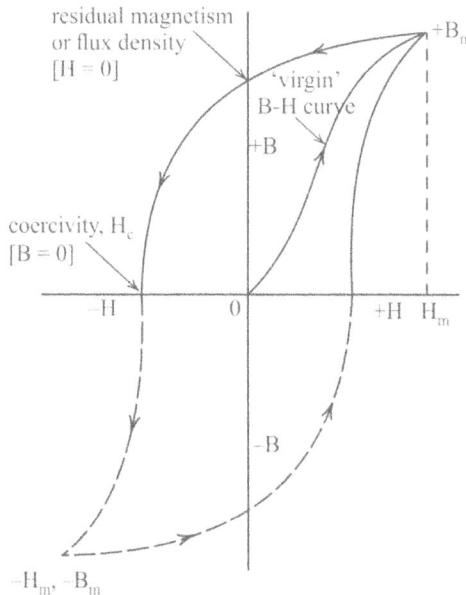

Fig.AP1.4 : Plotting of a hysteresis loop

Plotting a recoil loop

At any stage of plotting of the hysteresis loop, if it is desired to plot a major or minor recoil loop, the direction of current and hence the magnetising field is reversed at the given point on the B-H loop, increased in the positive direction up to the extreme point of the recoil loop and reversed back to the starting point on the loop. The process of plotting is then continued as before *without any break or pause*.

See illustrations later.

---

[6]If the process is followed correctly and smoothly, *without a pause*.

Measurement of J-H curve of a permanent magnet

It is often important to plot *intrinsic* magnetisation, J, with respect to magnetising field of a permanent magnet. Since J is related to the flux density in the material by

$$J = B - \mu_o H$$

the measurement of J is accomplished on the same hysteresigraph by using *both* the coils in the lower pole; one measuring the flux density as before whilst the other measuring the field strength, or rather the "airgap" flux density, μ_o H. For the purpose, the two coils, marked 3,3 in Fig.AP1.1, are connected in series opposition using switch 10, with only the left coil covering the test sample. Since both the coils are of identical design, the output to the flux meter, and hence to the plotter would be proportional to J. The plotting of the J-H curve and loop can then be carried out as in the case of a B-H curve.   However, for direct reading or measurement, a fresh calibration of the apparatus may be necessary.

Measurements on samples in powder form

Many magnetic materials that are available in powder form, sintered to provide permanent magnets of desired shape and size, for example ferrites, are to be tested in powder form to ascertain their 'basic' magnetic properties. For this, the available powder is tightly packed in a 'thin' *non-magnetic* annular ring of about 30 mm OD and 10 mm height, made of stainless steel or brass. The underneath of the composite sample is sealed by a very thin foil to retain the powder in flat-surface form[7]. The sample is then positioned between the poles as shown in Fig.AP1.5 and test performed as before.

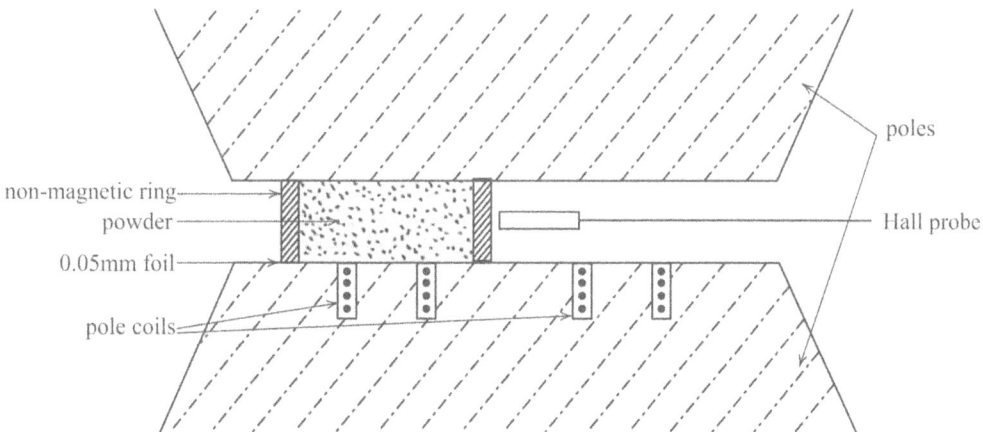

Fig.AP1.5 : B-H measurement on a sample in powder form

---

[7]If required, the powder may be mixed with a thermo-plastic resin as a bonding agent before being composted in the annular ring in order to prevent movement of the powder particles during measurement.

Calibration of the hysteresigraph set-up

When used with flexible, varying settings of controls on the H and B instruments, including the field regulator, and the scales of the X-Y recorder adjusted for measuring a variety of samples, a quick *overall* calibration of the hysteresigraph may be necessary for given settings.

A quite reliable calibration is achieved using a sample of pure nickel of the standard dimensions. This is feasible since pure nickel has a residual flux density of 0.621 T at 'room' temperature after magnetisation to well-saturated state. Thus, the nickel sample is tested with the given settings of the hysteresigraph and its 'B-H loop' recorded as shown in Fig.AP1.6. The B-H loops of other materials can then be related to the result for nickel as above.

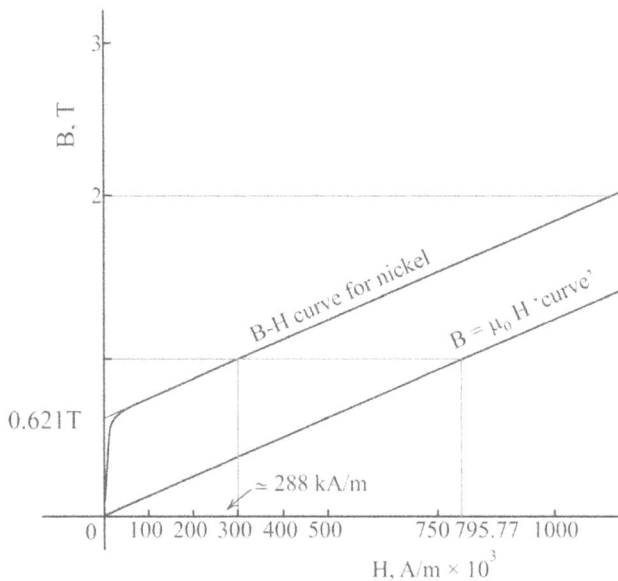

Fig.AP1.6 : "Saturation" characteristic of pure nickel

It is to be noted that the B-H *curve* for nickel rises sharply with only very slight increase of excitation, reaches 'saturation' and then follows almost the "air" magnetisation curve which when traced back without deviation yields the un-varying residual flux density, 0.621 T, as shown in Fig.AP1.6.

Miscellaneous Tests and Results

Three sample pieces, each of ALNICO, barium ferrite and samarian cobalt[8] were tested using the Magnet-Physic hysteresigraph following the procedure as above. The results are shown in Figs.AP1.7 through AP1.9.

---

[8]The former a brake magnet from a single-phase energy meter and the other a "standard" round, flat piece as supplied by manufacturer.

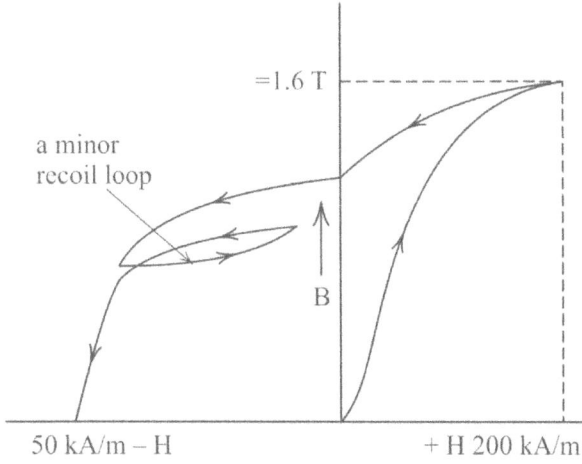

(a) virgin B-H curve and upper half B-H loop with one minor recoil loop

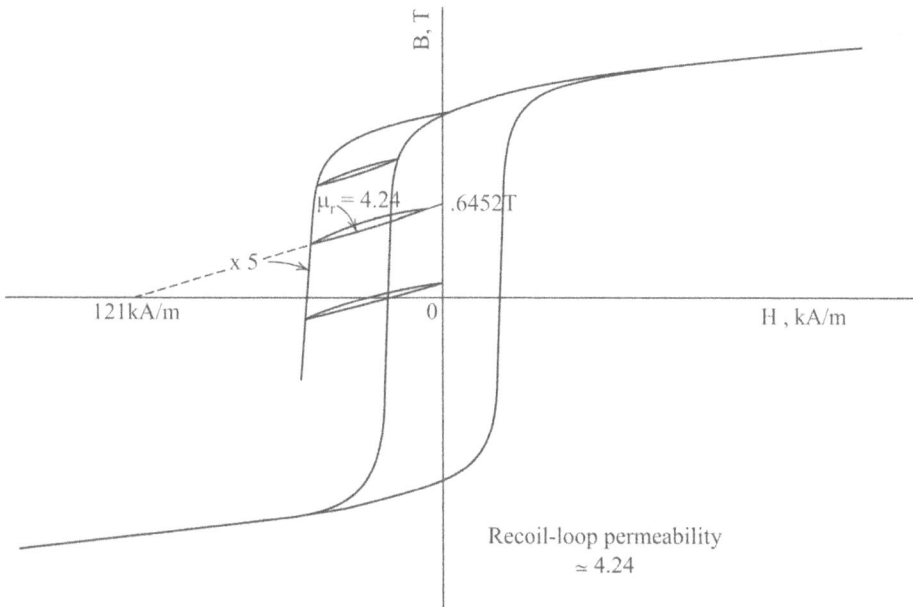

(b) B/H loop and recoil loops

Fig.AP1.7 : Magnetic characteristics of ALNICO

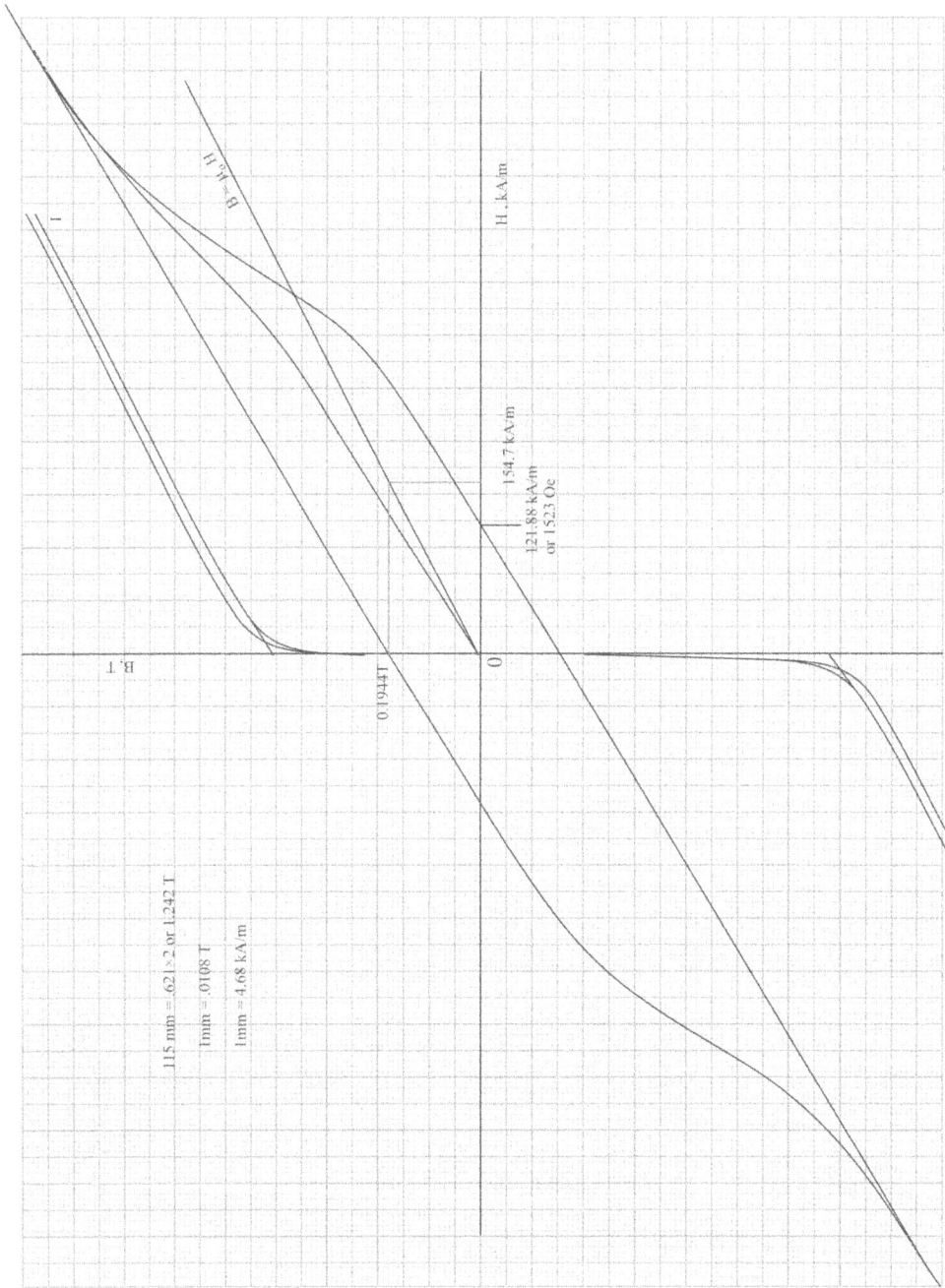

Fig.AP1.8 : Hysteresis loop of barium ferrite and characteristic of nickel

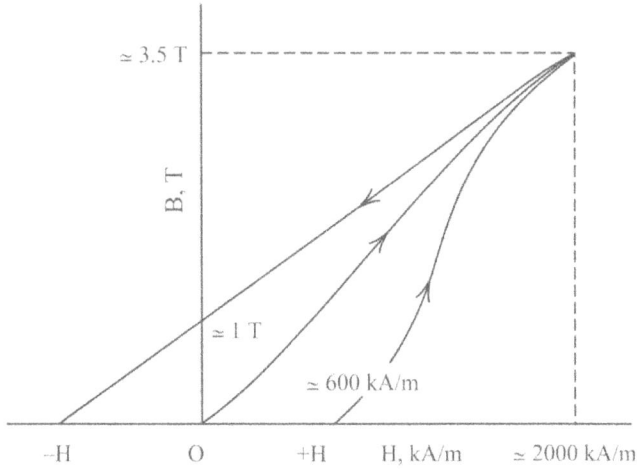

(a) virgin curve and upper-half B-H loop

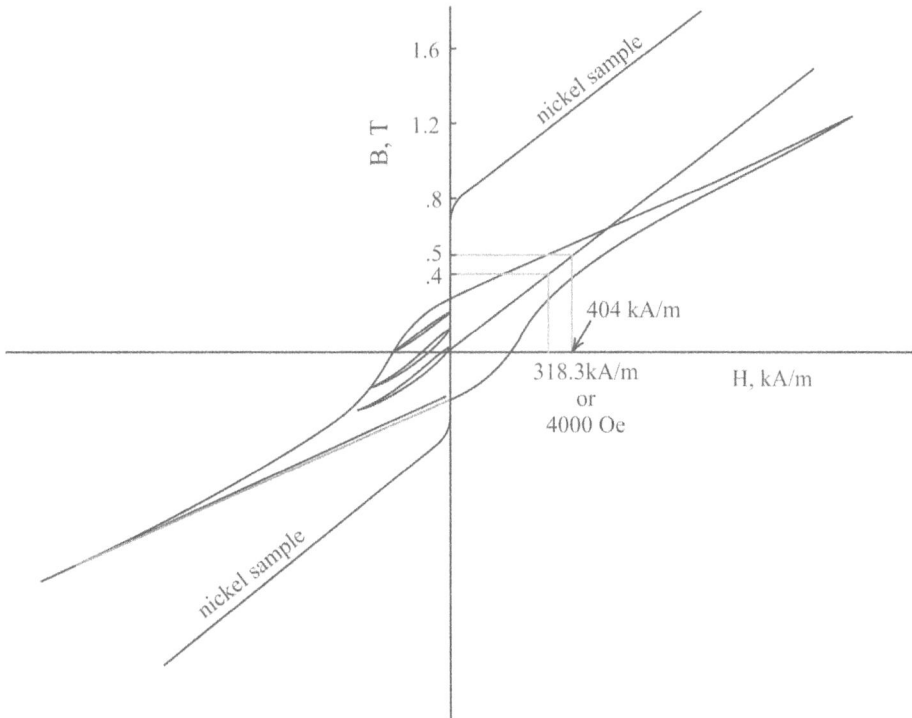

(b) B-H loop and recoil loops

Fig.AP1.9 : B/H characteristics of samarian cobalt

Measurement of Residual Flux of a Permanent Magnet

The magnetism or residual flux in a permanent magnet resides due to the magnetic polarisation of dipoles within the magnet. Under the influence of an externally applied field, the dipoles move, tending to 'align' or orient in the direction of the applied field when the latter is increased beyond the saturation value[9]. On removal of the field, however, the dipoles exhibit inherent hysteresis, leaving the magnet (partially) 'magnetised', this being identified as "residual" magnetism. The degree of residual flux or magnetism depends on a number of factors such as the composition of the magnet material, the process of manufacture (including any heat treatment, if applied) and magnetisation, and forms an important consideration for the design and application of a permanent magnet in a given device.

Method of measurement

To demonstrate the method[10], a small rectangular AlNiCo magnet (24mm × 12mm × 15mm), typically used as brake magnet in an energy meter, was used to measure its residual flux. Tested separately in the hysteresigraph, the B/H loop of the magnet showed B_r to be 1.13 T and H_c as 47 kA/m. The magnet was then assembled in a mild steel yoke as shown in Fig.AP1.10 (the assembly usually employed in the energy meter), leaving an airgap of 3.0 mm. A rectangular search coil of known number of turns, matching the area of cross-section of the magnet, but slightly loose so as to slide down the magnet, was wound and consolidated. The output of the search coil was measured on a calibrated flux meter as a result of change of flux linkages. The latter was effected by a quick movement of the search coil from around the magnet to the airgap and then quickly away from it to a 'sufficiently' long distance as illustrated in Fig.AP1.10.

Since the cross-section of the magnet is sufficiently large, the error due to the use of a loose-fitting search coil in the measurement may not be serious; else a 'correction' may be applied.

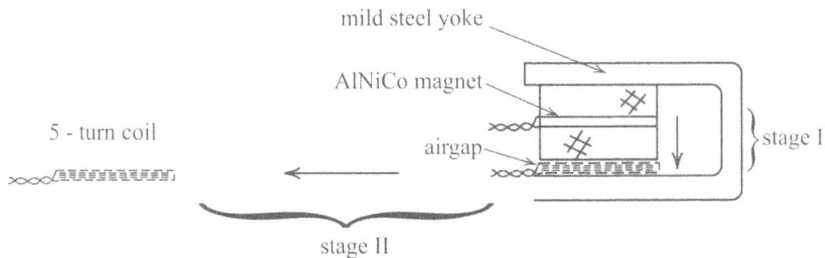

Fig.AP1.10 : Measurement of residual flux of a magnet

---

[9]See, for example,

R.M.Bozorth: Ferromagnetism (book), Van Nostrand Co. Inc., 1968.

[10]See, for example,

S,C. Bhargava: Measurement of residual flux in permanent magnets, Int. J. Elect. Enging. Edu., Vol. 20, 1983, pp 367-373.

Experimental Hysteresis Machine as a Permagraph[11]

Tests on samples in annulus form

Magnetic characteristics of vicalloy

The experimental hysteresis machine described in Part C comprised a rotor incorporating a vicalloy strip, 25 mm in width and 0.4 mm thick, bent to form an annulus, and a non-magnetic arbor within[12]. Whilst the magnetic properties of vicalloy – B-H curve and hysteresis loops, major and minor – were available from the manufacturer, it was deemed necessary to perform tests on the *actual* annulus, esp. as the latter was first formed in required shape and size, with necessary holes etc. drilled as described, followed by the recommended heat treatment of the annulus[13]. A conventional method[14] of making a "ring specimen" out of the annulus would not be practical on account of having to wind extremely large number of turns of the magnetising winding – a tedious process by itself – required to produce enormous magnetising field to drive the material into saturation, the material possessing relatively low relative permeability compared to usual ferromagnetic materials. Carrying of high to very high excitation current might also result in excessive heating and temperature rise of the annulus that might be detrimental to the very magnetic properties of the material.

Considering the above handicaps, an ingenious alternative was evolved[15] that was based on the use of the experimental hysteresis machine itself as a permagraph. A schematic of the arrangement is shown in Fig.AP1.11.

The two-pole field system of the machine was used to provide magnetisation of the annular sample positioned between, and in firm contact with, the poles. The flux through the annulus in the parts located on the interpolar axis would be uniformly distributed assuming the 'axial' length of the annulus being much larger than its thickness[16]. A five-

---

[11]See, for example,

S.C. Bhargava: A new technique for measurement of B/H characteristics of annular permanent magnets, Int. J. Elect. Enging Educ., Vol. 22, 1985, pp 221-27.

[12]Use of a very thin strip to form the annulus ensured that the hysteresis phenomenon was predominantly alternating.

[13]This consisted of heating the annulus in an electric furnace, soaked for several hours at 600 °C±2°C (ambient temperature being 17 °C). The ageing time was about 2 ½ hours and the annulus, held in a clamp, was allowed to cool down to room temperature in sill air. This led to development of 'full' magnetic properties in the annulus whilst making it "glass-hard" mechanically[48]. Quenching at high temperature (1,200 °C) as recommended in the literature[49] was found un-necessary.

[14]Foe example the method of reversals.

[15]By the author during his experimental research at the University of Aston, Birmingham.

[16]This being so in the present case with the length being 25.4 mm and thickness only 0.4 mm.

turn search coil[17] was wound, threaded through two 0.25 mm diameter holes in the annulus, symmetrical about its mid-width, using very fine, insulated (49 SWG) copper wire to measure the flux passing through the annulus corresponding to a given excitation or magnetising field; the flux density would then be derived from accurate knowledge of the annulus cross-section.

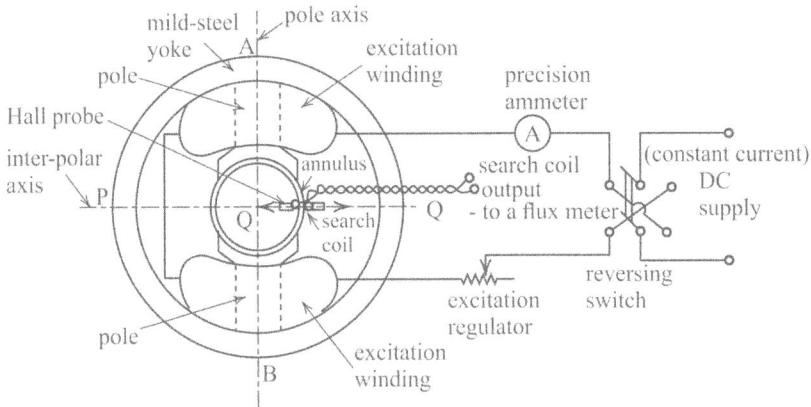

Fig.AP1.11 : Experimental hysteresis machine as a permagraph

Magnetising Force Inside the Annulus

Whilst the magnetising force inside the annulus could be obtained by placing a calibrated Hall probe by its side as in the case of measurements using the hysteresigraph, a more effective and accurate approach was adopted in the present procedure. The magnetising field was measured *externally* to the annulus on both sides using a transverse-type Hall probe, with its plane positioned at right angles to the pole axis as shown in Fig.AP1.11 and its sensing area held at mid-height of the annulus. With the exciting current stabilised at a selected value, the Hall probe is traversed away horizontally from as close to the annulus side as possible along the interpolar axis on both outside and inside the annulus as shown[18]. The corresponding flux density in the sample was measured using search coil output on a calibrated flux meter in the usual manner. As in the case of measurements with the hysteresigraph, the estimation of magnetising force inside the annulus was based on the property of continuity of tangential component of field on both sides of the annulus.

The Hall probe outputs along the axis at various excitations are plotted in Fig.AP1.12.

---

[17]See, for example,

S.C.Bhargaca: The use of search coils for magnetic measurements, Int. J. Elec. Enging Educ., Vol 19, pp. 45-52, 1982.

[18]See, for example,

S.C.Bhargava:A new technique for measurement of B/H characteristics of annular permanent magnets, *ibid.*, Vol 22, pp 221-227, 1985.

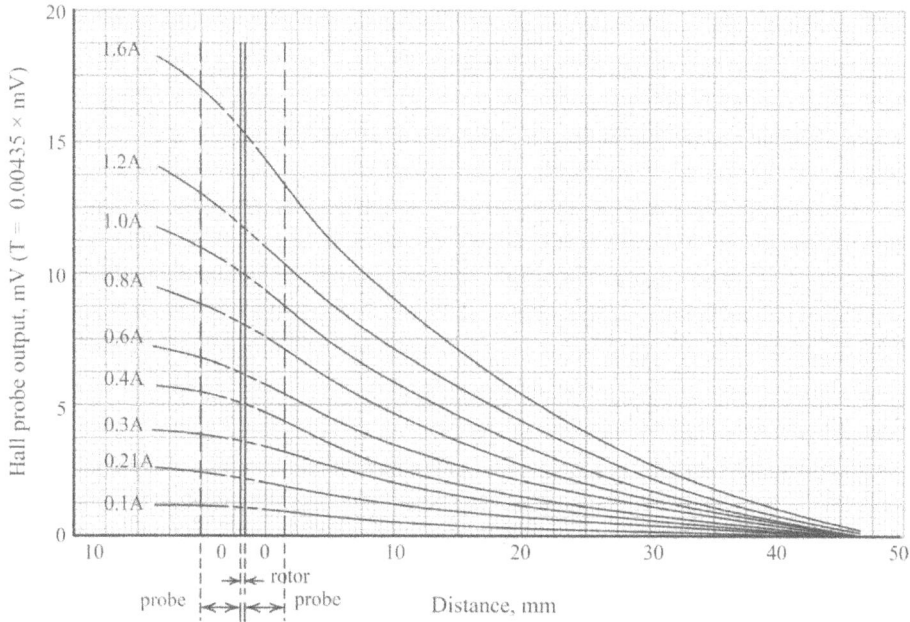

Fig.AP1.12 : "H" variation measured along interpolar axis using the Hall probe

From these plots, the value of magnetising force, H, *within* the annulus section is obtained by 'extrapolating' the measured H variation in air close to the internal and external surface.

Plotting B-H Curve

The B-H curve was then plotted from the values of H and B so obtained as listed in Table AP1.1. The curve is shown in Fig.AP1.13; also shown are the values obtained from the manufacturer[19] and those from a previous test.

Table AP1.1: B/H values for vicalloy

Excitation A	Extrapolated values of H for the rotor mV	Caculated values of H in the rotor A/m $\times 10^5$	Measured flux in the rotor µWb	Flux density value in the rotor tesla	Remarks
0.1	1.0	0.0346	0.75	0.03	B-H curve from the
0.21	2.22	0.0758	2.25	0.09	manufacturer and
0.3	3.78	0.131	4.25	0.17	previous work: as
0.44	5.38	0.186	12.0	0.48	available
0.6	6.5	0.224	22.0	0.88	Manufacturer: M/s
0.8	8.18	0.283	25.25	1.01	Telcon Metals Ltd.,
1.0	10.05	0.348	27.5	1.1	UK
1.2	11.9	0.442	28.0	1.12	
1.6	15.15	0.523	30.0	1.2	
2.0	19.5	0.675	31.25	1.25	

---

[19]M/s Telcon Metals Ltd., U K.

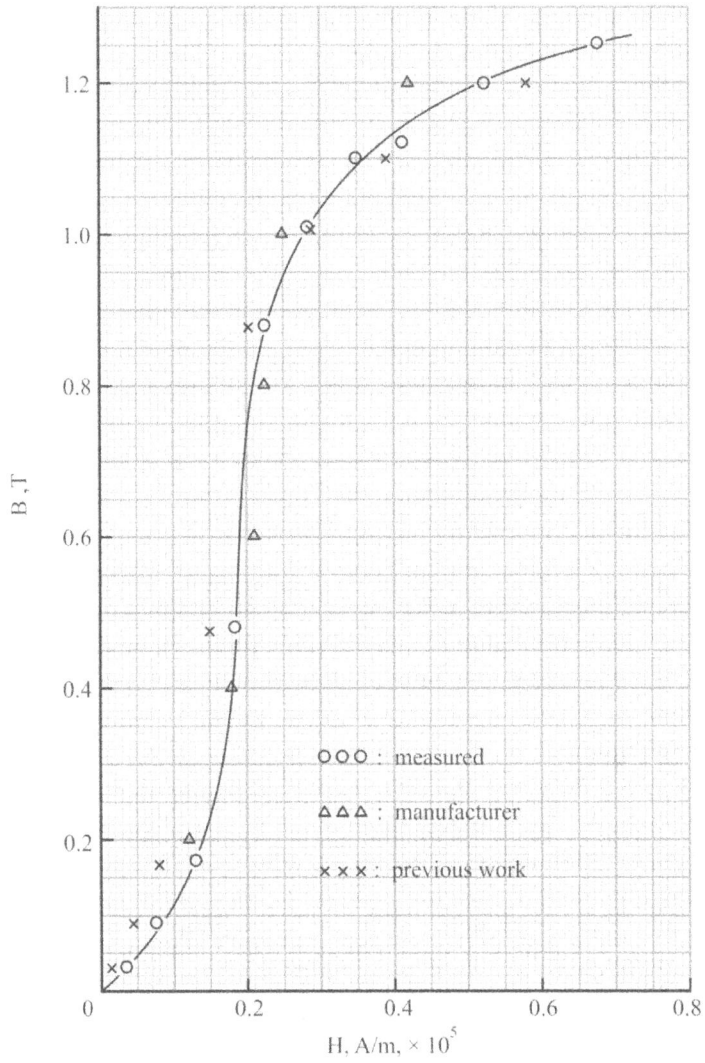

Fig.AP1.13 : B-H characteristic of vicalloy

Plotting B/H Loops

The above technique was extended to obtain hysteresis loops of vicalloy at various excitations. Eight experimental points were determined for different maximum values of excitation as described in Table AP1.2.

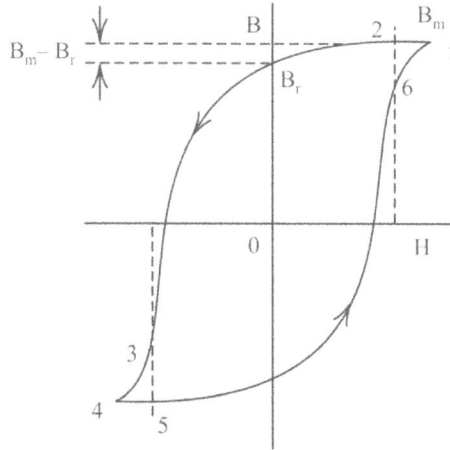

Fig.AP1.14 : Reference steps for plotting B/H loops

Table AP1.2: B/H values for hysteresis loops

	Point 1		Step 1 – 2		Step 2 – 3		Step 3 – 4		Step 4 – 5		Step 5 – 6		Maximum residual flux density,
Excitation A	H_m A/m × 10^5	B_m tesla	ΔH_1 A/m × 10^5	ΔB_1 tesla	ΔH_2 A/m × 10^5	ΔB_2 tesla	ΔH_3 A/m × 10^5	ΔB_3 tesla	ΔH_4 A/m × 10^5	ΔB_4 tesla	ΔH_5 A/m × 10^5	ΔB_5 tesla	$B_m - B_{res}$. (obtained separetely) tesla
0.1	0.0.35	0.03	0.01	0.0	0.05	0.04	0.01	0.0	0.01	0.0	0.05	0.06	
0.2	0.079	0.1	0.022	0.0	0.114	0.12	0.022	0.02	0.022	0.0	0.114	0.12	0.05
0.3	0.134	0.19	0.037	0.02	0.194	0.22	0.037	0.06	0.037	0.02	0.194	0.22	0.1
0.4	0.176	0.41	0.39	0.02	0.274	0.5	0.039	0.3	0.039	0.02	0.274	0.44	0.14
0.6	0.226	0.9	0.043	0.02	0.366	1.5	0.043	0.26	0.043	0.0	0.366	1.5	0.14
0.8	0.288	1.05	0.06	0.02	0.456	1.9	0.06	0.14	0.06	0.02	0.456	1.9	0.16
1.0	0.352	1.1	0.075	0.04	0.554	2.1	0.075	0.1	0.075	0.0	0.554	2.1	0.18
1.2	0.425	1.15	0.094	0.04	0.662	2.2	0.094	0.04	0.094	0.04	0.662	2.2	0.2
1.6	0.545	1.2	0.11	0.06	0.87	2.3	0.11	0.06	0.11	0.02	0.87	2.3	0.26
2.0	0.675	1.25	0.15	0.06	1.05	2.4	0.15	0.04	0.15	0.0	1.05	2.4	0.28

The loops are plotted in Fig.AP1.15.

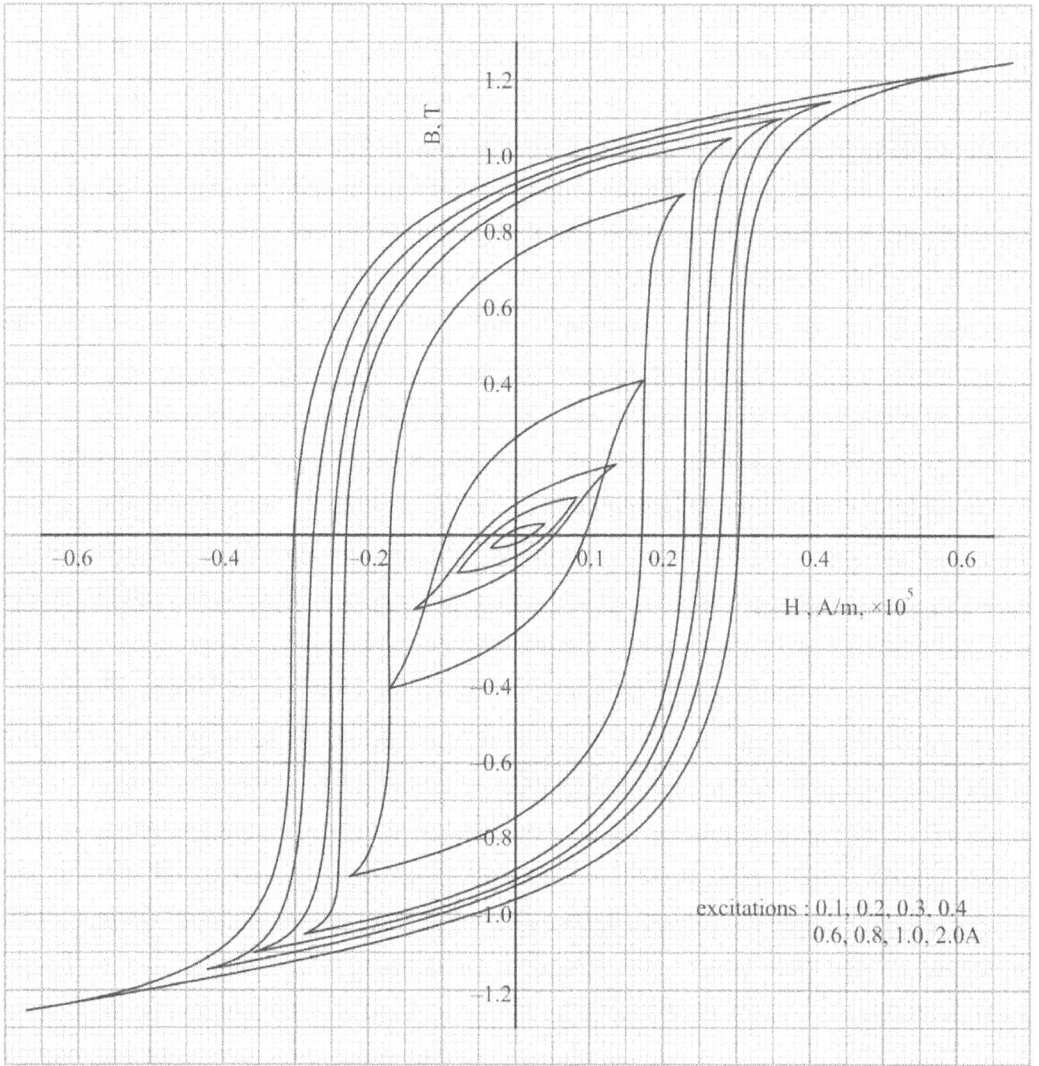

Fig.AP1.15: Hysteresis loops for vicalloy

Alternating Hysteresis Loss

An alternating hysteresis loss curve was derived from the hysteresis loops by measuring the area of each curve and plotted against maximum flux density values in the rotor as shown in Fig.AP1.16.

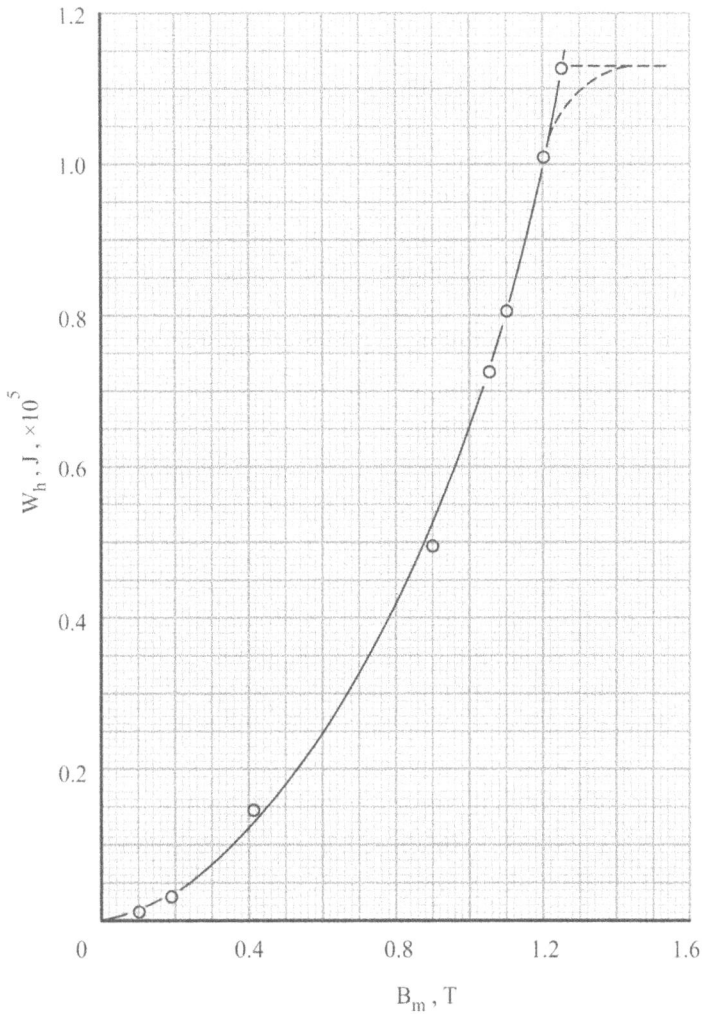

Fig.AP1.16 : Alternating hysteresis loss in vicalloly

The upper part of the curve shown dotted is the approximation for the loss reaching a final steady value corresponding to intense saturation in vicalloy: as the mmf is increased, the maximum flux density still increases[48], but there is little increase in hysteresis loss.

2

Digital computer program for the calculation of flux density and power loss

---

The first part of the program, originally written in ALGOL60 in the form of PROCEDURE, deals with harmonic analysis of the measured waveforms; the second part is the main program for calculation of flux density and power flow.

The object of using an elaborate, unique method for harmonic analysis of B_r or B_θ waveforms was found imperative owing to very sharp peaks characterising the waveforms, with relatively flat portions under the poles. With the usual methods, based for example on calculus of variation, the resolution of such waveforms into higher-order harmonics, responsible for the peaks, would require large number of equally-spaced ordinates and even then the results may be erroneous[1]

The Procedure

The method in the present research is centred on 'integrating by parts' pre-selected, consecutive portions of the waveform to evaluate Fourier coefficients for each harmonic. Commencing at the beginning of the waveform under the pole centre (or the interpolar axis), each portion could be described by three successive points as indicated in Fig.AP2.1, joined to provide either a 'liner' or 'quadratic' variation or 'fit', the latter being more elaborate to make up for smooth variation of that portion of the waveform[2].

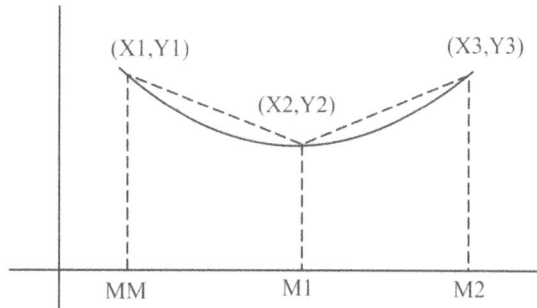

Fig.AP2.1 : A portion of waveform described by three consecutive points

---

[1]This may be obviated to an extent by employing now commonly available methods of digital processing of data.

[2]See, for example,

S.C.Bhargava: A unique method for Fourier analysis of distorted periodic waveforms encountered in engineering problems, BHEL Journal, Vol. 7, No. 1, 1986, pp 35-41.

Clearly, each type of fit would depend on the relative location of the three points. The mathematical check for these are included in program POLYFIT to decide the type of fit, and label (or 'write') the same, as illustrated in the flow diagram of Fig.AP2.2. Taken each fit at a time in succession would cover the entire waveform and when analysed would provide the necessary coefficients for the sine and cosine terms.

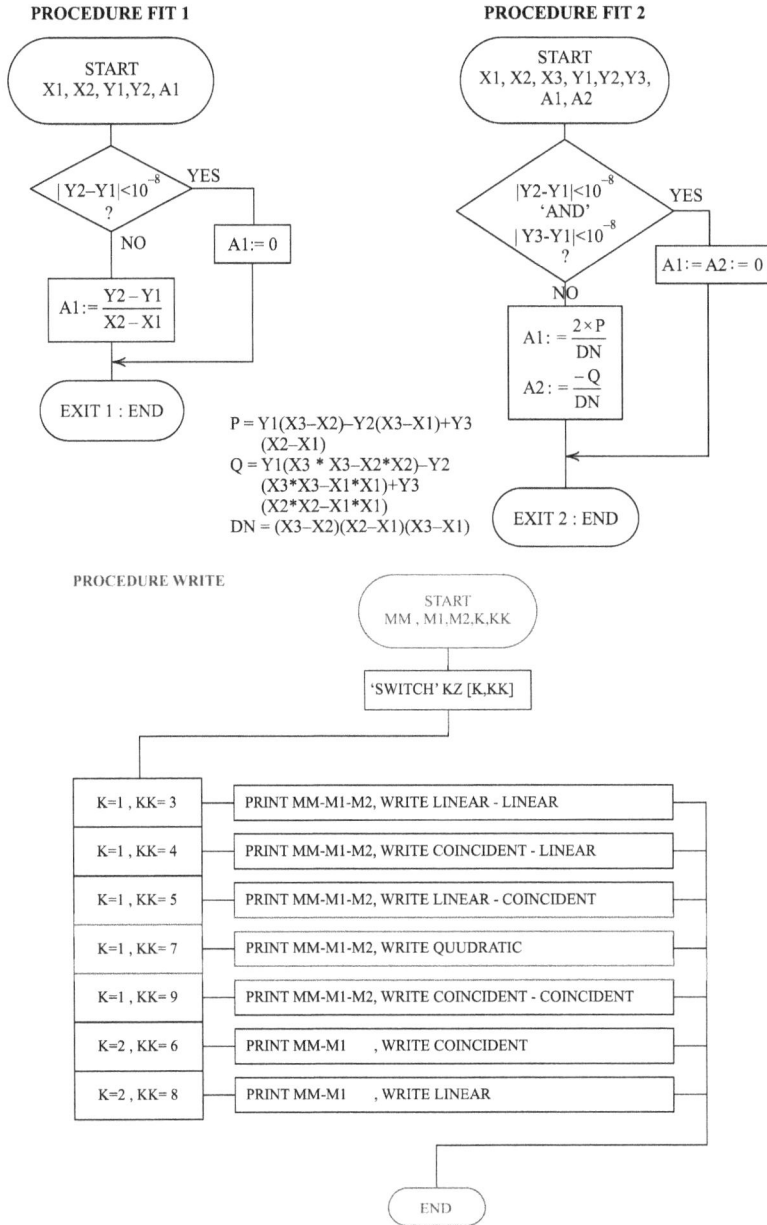

Fig.AP2.2 : Flow diagram for Procedure POLYFIT and WRITE

START
X1,X2, . . . Y1,Y2, . . . M

$AV:=\sum|Y1|/(M-1)$, $Y_X:=Y_X/AV$, T:=1

MM:=1

M1:=MM+1, M2:=MM+2

M1>M ? — YES

NO

* P

M2>M ? — NO

X3:=X[M2];Y3:=Y[M2]
X13:=X3-X1;X23:=X3-X2

YES

X12<10⁻⁶ ? — YES

NO

WRITE(MM,M1,M2,2,8)

WRITE(MM,M1,M2,2,6)

LAST:=1

$X12<10^{-6}$ 'AND' $X23<10^{-6}$? — YES

WRITE(MM,M1,M2,1,9)

NO

Y23:=X12/X13

$X12<10^{-6}$? — YES

WRITE(MM,M1,M2,1,4)

NO

$X23<10^{-6}$? — YES

WRITE(MM,M1,M2,1,5)

NO

VERT:=1

Y23>0.8 'OR' Y23<0.2 ? —

WRITE(MM,M1,M2,1,3)

NO

** X12:=f₁(X,Y)
X13:=f₂(X,Y)

X13>X12 ? — YES

WRITE(MM,M1,M2,1,3)

NO

WRITE(MM,M1,M2,1,7)

FIT2(X1,X2,X3,Y1,Y2,Y3,A1,A2)
SCN[T,1]:=SCN[T,1]-(A1+A2*X1)
SCN[T,2]:=SCN[T,2]-A2
SCN[T+1,1]:=A1+A2*X3;SCN[T+1,2]:=A2
X[T]:=X1;X[T+1]:=X3

FIT1(X1,X2,Y1,Y2,A1)
SCN[T,1]:=SCN[T,1]-A1
SCN [T+1,1]:=A1;SCN[T+1,2]:=0
X[T]:=X1;X[T+1]:=X2

FIT1(X2,X3,Y2,Y3,A1)
SCN[T,1]:=SCN[T,1]-A1
SCN[T+1,1]:=A1;SCN[T+1,2]:=0
X[T]:=X2;X[T+1]:=X3

T=1 ? — YES

NO

Y[1]:=-Y1

Y[T]:=Y[T]-Y1

Y[T+1]:=Y3
T:=T+1

T=1 ? — YES

NO

Y[1]:=-Y1

Y[T]:=Y[T]-Y1

Y[T+1]:=Y2
T:=T+1

T=1 ? — YES

NO

Y[1]:=-Y2

Y[T]:=Y[T]-Y2

Y[T+1]:=Y3
T:=T+1

LAST=1 ? — YES

NO

VERT=1 ? — YES

NO

VERT:=0

TF4:END

MM:=MM+2 — YES

LAST=0 ?

NO

TFS:END POLYFIT

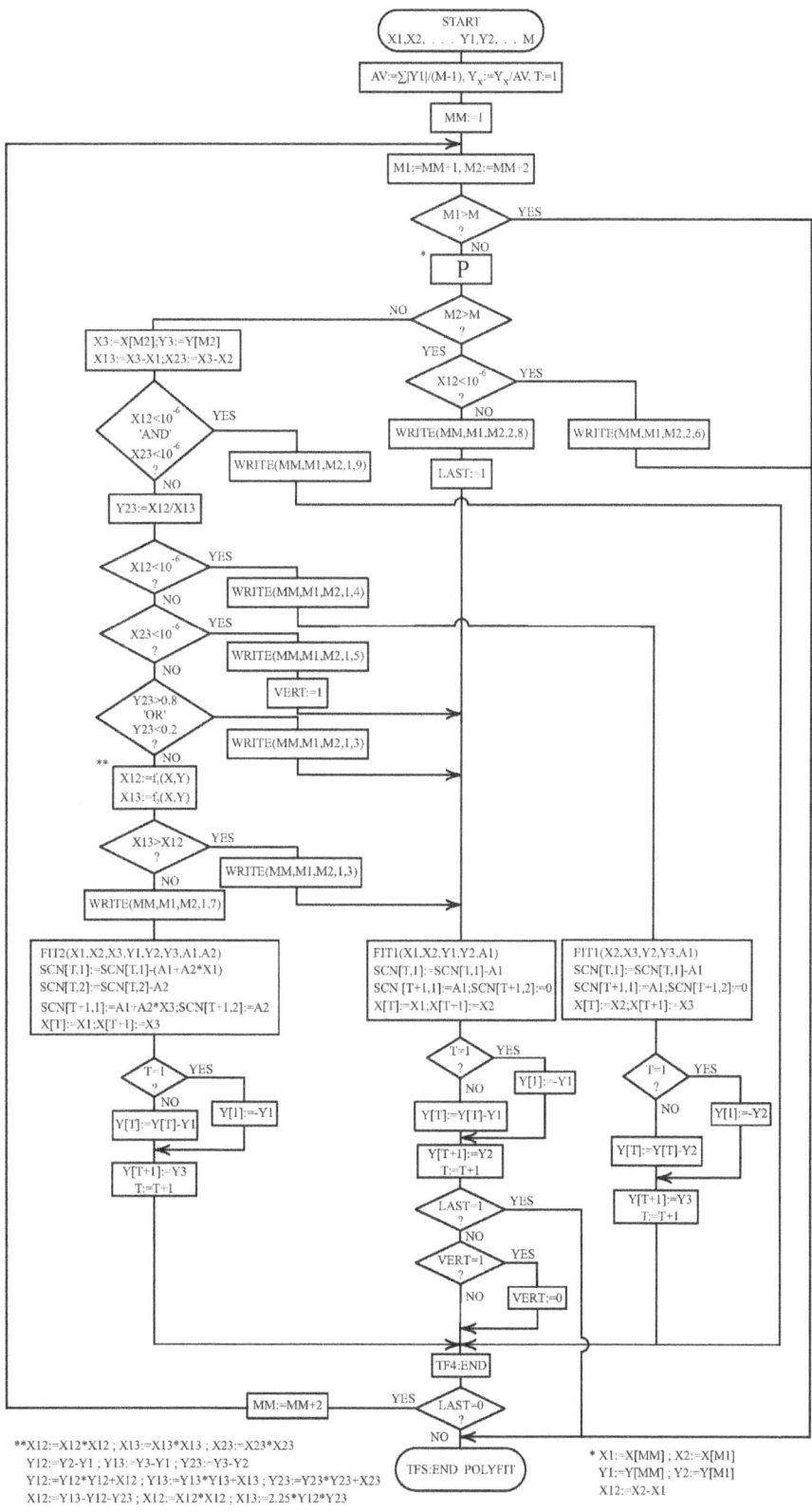

**X12:=X12*X12 ; X13:=X13*X13 ; X23:=X23*X23
Y12:=Y2-Y1 ; Y13:=Y3-Y1 ; Y23:=Y3-Y2
Y12:=Y12*Y12+X12 ; Y13:=Y13*Y13+X13 ; Y23:=Y23*Y23+X23
X12:=Y13-Y12-Y23 ; X12:=X12*X12 ; X13:=2.25*Y12*Y23

* X1:=X[MM] ; X2:=X[M1]
Y1:=Y[MM] ; Y2:=Y[M1]
X12:=X2-X1

Insert AP1 : Flow diagram of PROCEDURE POLYFIT

The Complete Program

The flow diagrams for the complete program are given in Insert AP1 and Fig.AP2.3, followed by the listing of the complete program.

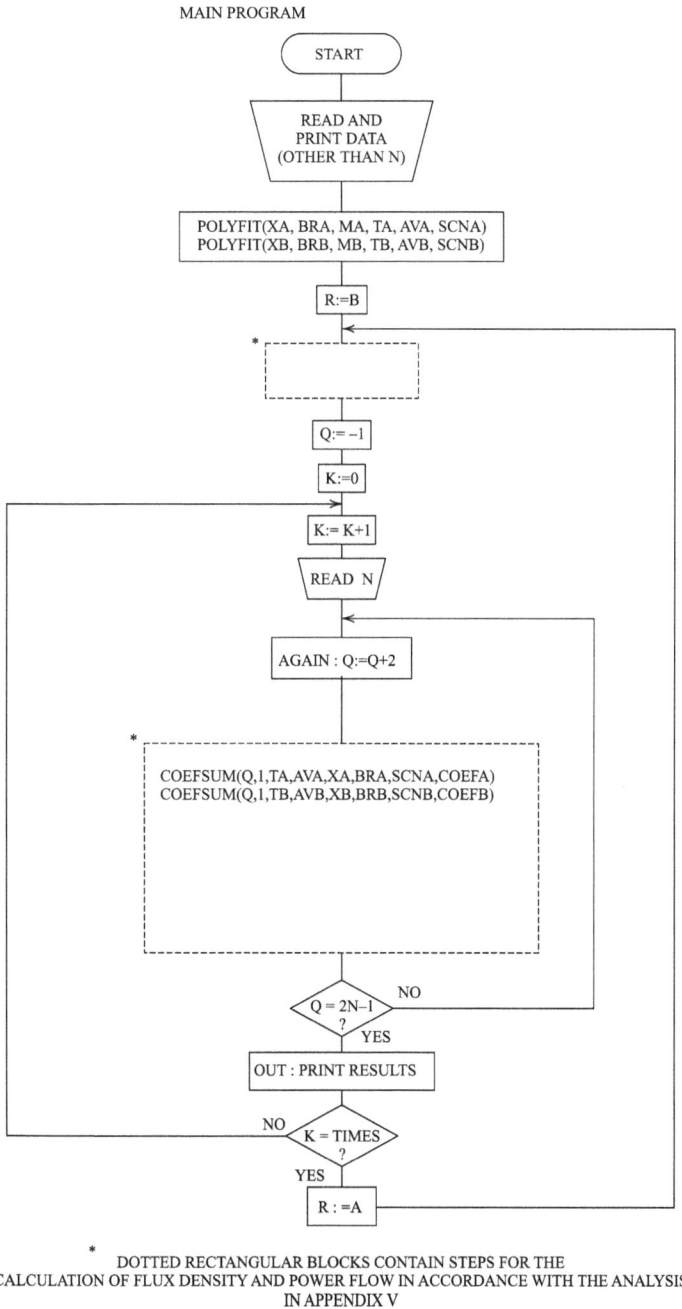

MAIN PROGRAM

START

READ AND
PRINT DATA
(OTHER THAN N)

POLYFIT(XA, BRA, MA, TA, AVA, SCNA)
POLYFIT(XB, BRB, MB, TB, AVB, SCNB)

R:=B

*

Q:= −1

K:=0

K:= K+1

READ N

AGAIN : Q:=Q+2

*

COEFSUM(Q,1,TA,AVA,XA,BRA,SCNA,COEFA)
COEFSUM(Q,1,TB,AVB,XB,BRB,SCNB,COEFB)

Q = 2N−1
?   NO
YES

OUT : PRINT RESULTS

NO   K = TIMES
?
YES

R : =A

*   DOTTED RECTANGULAR BLOCKS CONTAIN STEPS FOR THE
CALCULATION OF FLUX DENSITY AND POWER FLOW IN ACCORDANCE WITH THE ANALYSIS
IN APPENDIX V

Fig.AP2.3 : Flow diagrams of MAIN PROGRAM

PROGRAM LISTING in ALGOL 60

```
'BEGIN' 'COMMENT' BHARGAVA EEPSF088 TRIGFIT HARMONIC ANALYSIS (ODD HARMONICS ONLY);   HAR 0000
'PROCEDURE' POLAR(X,Y,R,PHI);                                                         POL 100
'VALUE' X,Y; 'REAL' X,Y,R,PHI;                                                        POL 200
'BEGIN' R:= SQRT(X*X + Y*Y);                                                          POL 300
        PHI:='IF' X=0 'THEN' 90*SIGN(Y) 'ELSE' 57.295779513*ARCTAN(Y/X) +            POL 400
        ('IF' X 'GE' 0 'THEN' 0 'ELSE' 'IF' Y 'GE' 0 'THEN' 180 'ELSE' -180);        POL 500
 'END' POLAR ;                                                                        POL 600
'PROCEDURE' POLYFIT(X,Y,M,T,AV,SCN);                                                  PFT 100
'VALUE' M; 'REAL' AV; 'INTEGER' M,T;   'ARRAY' X,Y,SCN;                               PFT 200
'BEGIN' 'REAL' A1,A2,X1,X2,X3,X12,X13,X23,Y1,Y2,Y3,Y12,Y13,Y23;                       PFT 300
        'INTEGER' M1,M2,MM,N5,N6,NN,P,LAST,VERT;                                      PFT 400
        'PROCEDURE' FIT1(X1,X2,Y1,Y2,A1);                                            PFT 500
        'VALUE' X1,X2,Y1,Y2;   'REAL' X1,X2,Y1,Y2,A1;                                PFT 600
        'BEGIN' 'IF' ABS(Y2-Y1) < 1.0 & -8 'THEN'                                    PFT 700
                'BEGIN' A1:=0; 'GOTO' EXIT1; 'END';                                  PFT 800
                A1:= (Y2-Y1)/(X2-X1);                                                PFT 900
           EXIT1 : 'END';                                                            PFT 1000
        'PROCEDURE' FIT2(X1,X2,X3,Y1,Y2,Y3,A1,A2);                                   PFT 1100
        'VALUE' X1,X2,X3,Y1,Y2,Y3;   'REAL' X1,X2,X3,Y1,Y2,Y3,A1,A2;                 PFT 1200
        'BEGIN' 'REAL' N1,N2,N3,DN,Z1,Z2,Z3;                                         PFT 1300
                'IF' ABS(Y2-Y1) < 1.0 & -8 'AND' ABS(Y3-Y1) < 1.0 & -8 'THEN'        PFT 1400
                'BEGIN' A1:=A2:=0; 'GOTO' EXIT2; 'END';                              PFT 1500
                N1:=X1+X2; N2:=X1+X3; N3:=X2+X3; Z2:=X2-X1; Z3:=X3-X1;               PFT 1600
                Z1:=X3-X2; DN:=1/(Z1*Z2*Z3); Z1:=Z1*Y1; A2:=Z3*Y2;                   PFT 1700
                Z3:=Z2*Y3; Z2:=A2; A2:=2*(Z1-Z2+Z3)*DN;                             PFT 1800
                Z1:=Z1*N3; Z2:=Z2*N2; Z3:=Z3*N1; A1:=-(Z1-Z2+Z3)*DN;                PFT 1900
           EXIT2 : 'END';                                                            PFT 2000
        'PROCEDURE' WRITE(MM,M1,M2,K,KK);                                            PFT 2100
        'VALUE' MM,M1,M2,K,KK;   'INTEGER' MM,M1,M2,K,KK;                            PFT 2200
        'BEGIN' 'SWITCH' KZ:= K1,K2,K3,K4,K5,K6,K7,K8,K9,K10;                        PFT 2300
                NEWLINE(1); 'GOTO' KZ[K];                                            PFT 2400
           K1 : PRINT(MM,3,0); WRITETEXT('(' - ')'); PRINT(M1,3,0);                  PFT 2500
                WRITETEXT('(' -')'); PRINT(M2,3,0); 'GOTO' KZ[KK];                   PFT 2600
           K2 : SPACE(4); PRINT(MM,3,0);WRITETEXT('(' -')');                         PFT 2700
                PRINT(M1,3,0); SPACE(3); 'GOTO' KZ[KK];                              PFT 2800
           K3 : WRITETEXT('('('14S')'LINEAR'('S')'-'('S')'LINEAR')');                PFT 2900
                'GOTO' K10;                                                          PFT 3000
           K4 : WRITETEXT('('('10S')'COINCIDENT'('S')'-'('S')'                       PFT 3100
                LINEAR')'); 'GOTO' K10;                                              PFT 3200
           K5 : WRITETEXT('('('14S')'LINEAR'('S')'-'('S')'                           PFT 3300
                COINCIDENT')'); 'GOTO' K10;                                          PFT 3400
           K6 : WRITETEXT('('('16S')'COINCIDENT')'); 'GOTO' K10;                     PFT 3500
           K7 : WRITETEXT('('('17S')'QUADRATIC')'); 'GOTO' K10;                      PFT 3600
           K8 : WRITETEXT('('('18S')'LINEAR')'); 'GOTO' K10;                         PFT 3700
           K9 : WRITETEXT('('('10S')'COINCIDENT'('S')'-'('S')'                       PFT 3800
                COINCIDENT')');                                                      PFT 3900
           K10 : 'END';                                                              PFT 4000
        AV:=0; 'FOR' MM:= 1 'STEP' 1 'UNTIL' M 'DO' AV:=AV+ABS(Y[MM]);               PFT 4100
        AV:=AV/(M-1); 'FOR' MM:= 1 'STEP' 1 'UNTIL' M 'DO'                           PFT 4200
        Y[MM]:=Y[MM]/AV; T:=1; SCN[1,1]:=SCN[1,2]:=0;                                PFT 4300
        WRITETEXT('('('3C15S')'CURVE'('S')'FITTING'('S')'ANALYSIS                    PFT 4400
        '('2C7S')'POINTS'('26S')'TYPE'('1C')'')'); LAST:=VERT:=0;                    PFT 4500
        'FOR' MM:= 1,MM+2 'WHILE' LAST=0 'DO'                                        PFT 4600
```

```
'BEGIN' M1:=MM+1;  M2:=MM+2;  'IF' M1>M 'THEN' 'GOTO' TF5;            PFT 4700
       X1:=X[MM]; X2:=X[M1]; Y1:=Y[MM]; Y2:=Y[M1]; X12:=X2-X1;         PFT 4800
       'IF' M2>M 'THEN' 'BEGIN' 'IF' X12<1.0 &-6 'THEN' 'BEGIN'       PFT 4900
       WRITE(MM,M1,M2,2,6); 'GOTO' TF5; 'END'; WRITE(MM,M1,M2,2,8);   PFT 5000
       LAST:=1; 'GOTO' TF1; 'END'; X3:=X[M2]; Y3:= Y[M2]; X13:=X3-X1;  PFT 5100
       X23:=X3-X2; 'IF' X12<1.0 &-6 'AND' X23<1.0 &-6 'THEN'          PFT 5200
       'BEGIN' WRITE(MM,M1,M2,1,9); 'GOTO' TF4; 'END';                 PFT 5300
       Y23:=X12/X13; 'IF' X12<1.0 &-6 'THEN' 'BEGIN' WRITE(MM,M1,M2,1,4); PFT 5400
       'GOTO' TF2; 'END'; 'IF' X23<1.0 &-6 'THEN' 'BEGIN' WRITE(MM,M1,M2,1,5)PFT 5500
       ; VERT:=1; 'GOTO' TF1; 'END'; 'IF' Y23>0.8 'OR' Y23<0.2 'THEN'  PFT 5600
       'BEGIN' WRITE(MM,M1,M2,1,3);  'GOTO' TF1; 'END'; X12:=X12*X12;  PFT 5700
       X13:=X13*X13; X23:=X23*X23; Y12:=Y2-Y1; Y13:=Y3-Y1; Y23:=Y3-Y2; PFT 5800
       Y12:=Y12*Y12+X12; Y13:=Y13*Y13+X13; Y23:=Y23*Y23+X23;          PFT 5900
       X12:=Y13-Y12-Y23; X12:=X12*X12; X13:=2.25*Y12*Y23;             PFT 6000
       'IF' X13>X12 'THEN' 'BEGIN' WRITE(MM,M1,M2,1,3); 'GOTO' TF1; 'END' PFT 6100
       'ELSE' 'BEGIN' WRITE(MM,M1,M2,1,7); 'GOTO' TF3; 'END';          PFT 6200
  TF1 : FIT1(X1,X2,Y1,Y2,A1); SCN[T,1]:=SCN[T,1]-A1;                   PFT 6300
       SCN[T+1,2]:=0; SCN[T+1,1]:=A1; X[T]:=X1; X[T+1]:=X2;            PFT 6400
       'IF' T=1 'THEN' Y[1]:= -Y1; 'ELSE' Y[T]:=Y[T]-Y1; Y[T+1]:=Y2; T:=T+1; PFT 6500
       'IF' LAST = 1 'THEN' 'GOTO' TF5; 'IF' VERT = 1 'THEN' 'BEGIN'   PFT 6600
       VERT:=0; 'GOTO' TF4; 'END';                                     PFT 6700
  TF2 : FIT1(X2,X3,Y2,Y3,A1); SCN[T,1]:=SCN[T,1]-A1;                   PFT 6800
       SCN[T+1,2]:=0; SCN[T+1,1]:=A1; X[T]:=X2; X[T+1]:=X3;            PFT 6900
       'IF' T=1; 'THEN' Y[1]:= -Y2; 'ELSE' Y[T]:=Y[T]-Y2; Y[T+1]:=Y3;  PFT 7000
       T:=T+1;  'GOTO' TF4;                                            PFT 7100
  TF3 : FIT2(X1,X2,X3,Y1,Y2,Y3,A1,A2); SCN[T,1]:=SCN[T,1]-(A1+A2*X1);  PFT 7200
       SCN[T+1,1]:=A1+A2*X3; SCN[T,2]:=SCN[T,2]- A2; SCN[T+1,2]:=A2;   PFT 7300
       X[T]:=X1; X[T+1]:=X3; 'IF' T=1 'THEN' Y[1]:= -Y1 'ELSE' Y[T]:=Y[T]-Y1; PFT 7400
       Y[T+1]:=Y3; T:=T+1;                                            PFT 7500
  TF4 : 'END';                                                         PFT 7600
TF5 : 'END' POLYFIT ;                                                  PFT 7700
'PROCEDURE' COEFSUM(HS,N,T,AV,X,Y,SCN,COEF);                           CFS 100
'VALUE' HS,N,T,AV; 'REAL' AV; 'INTEGER' HS,N,T; 'ARRAY' X,Y,SCN,COEF;  CFS 200
'BEGIN' 'REAL' A1,A2,X2,X3,X12,X13,Y12,Y13,Y23;  'INTEGER' MM,N5;      CFS 300
       'FOR' N5:= 1 'STEP' 1 'UNTIL' N 'DO'                            CFS 400
       'BEGIN' X12:=2*N5-2.0+HS; Y12:=1/X12; Y13:=Y12*Y12; Y23:=Y12*Y13; CFS 500
       COEF[1,N5]:=COEF[2,N5]:=0;                                      CFS 600
       'FOR' MM:= 1 'STEP' 1 'UNTIL' T 'DO' 'BEGIN' X13:=X12*X[MM];    CFS 700
       X2:=SIN(X13); X3:=COS(X13); A1:=Y[MM]*Y12 - SCN[MM,2]*Y23;      CFS 800
       A2:=SCN[MM,1]*Y13; COEF[1,N5]:=COEF[1,N5]- A1*X3+A2*X2;         CFS 900
       COEF[2,N5]:=COEF[2,N5] + A1*X2+A2*X3;  'END';                   CFS 1000
              COEF[1,N5]:= 0.6366199724*AV*COEF[1,N5];                 CFS 1100
              COEF[2,N5]:= 0.6366199724*AV*COEF[2,N5];                 CFS 1200
       'END' ;                                                         CFS 1300
'END' COEFSUM ;                                                        CFS 1400
```

```
'COMMENT' PROPER PROGRAMME STARTS HERE ;                                    HAR  100
'INTEGER' N,P,Q,MA,MB,NA,TA,TB,K, TIMES ;                                   HAR  200
'REAL' AVA,AVB,A,B,R,B3,B4,B5,B6,RA,RA1,RB,RB1,ML,AC,RI,PN1,PN2,QN1,QN2,PN,QN, HAR 300
       LMA,LMB,B7,RN,SN,PVI,PI,OM,PHI1,PHI2,MU,MUI,BA,Y,Z,C,PHIP1,PHIP2,T,NR, HAR 400
       GX1,GX2 ;                                                            HAR  500
       TIMES:= READ; NA:=READ; NR:=READ; GX1:=READ; GX2:=READ; T:=READ;     HAR  600
       MA:=READ; MB:=READ; LMA:=READ; LMB:=READ;                           HAR  700
'COMMENT' T IS TIME IN SECONDS FOR NR REVOLUTIONS OF THE FIELD SYSTEM MEASURED BY HAR 800
       THE ELECTRONIC TIMER. GX1 AND GX2 ARE GALVANOMETER CONSTANTS (MV/MM) ; HAR 900
       PI:= 3.14159265359; LMA:= PI/LMA; LMB:= PI/LMB;                     HAR 1000
       OM:= 2.0*PI*(NR/T); MU:= 4.0*PI*10 ↑(-7); MUI:= 1/MU; Y:= 1/(25.4*OM); HAR 1100
'BEGIN' 'ARRAY' XA,BRA[1:MA], XB,BRB[1:MB], COEFA,COEFB[1:2,1:1],           HAR 1200
       SCNA[1:MA,1:2], SCNB[1:MB,1:2], AM,AN,BT,CC,CS,BR,BX,PA,PHI,PHIP[1:NA]; HAR 1300
       WRITETEXT('('('2C')' DATA%FROM%RADIAL%FLUX%DENSITY%WAVEFORM%-%THE%FULL% HAR 1400
       PITCH%COIL'('30S')' GX1%(MV/MM)%= ')'); PRINT(GX1,2,8);             HAR 1500
              WRITETEXT('('('2C2S')'MA'('16S')'XA'('17S')'BRA'('13S')'     HAR 1600
              BRA%IN%MV '('2C')'')');                                      HAR 1700
              'FOR' P:= 1 'STEP' 1 'UNTIL' MA 'DO'                         HAR 1800
              'BEGIN' NEWLINE(1); PRINT(P,3,0); SPACE(10); XA[P]:=READ;    HAR 1900
                     PRINT(XA[P],3,2); SPACE(10); BRA[P]:=READ; PRINT(BRA[P],3,2); HAR 2000
                     SPACE(9); BRA[P]:= BRA[P]* GX1; PRINT(BRA[P],3,4);    HAR 2100
                     XA[P]:= XA[P]* LMA;                         'END';     HAR 2200
       PAPERTHROW ;                                                        HAR 2300
              WRITETEXT('('('2C')' DATA%FROM%THE%'DIFFERENCE'%WAVEFORM '('30S')' HAR 2400
              GX2%(MV/MM)%= ')') ; PRINT(GX2,2,8);                         HAR 2500
              WRITETEXT('('('2C2S')'MB'('16S')'XB'('17S')'BRB'('13S')'     HAR 2600
              BRB%IN%MV '('2C')'')') ;                                     HAR 2700
              'FOR' P:= 1 'STEP' 1 'UNTIL' MB 'DO'                         HAR 2800
              'BEGIN' NEWLINE(1); PRINT(P,3,0); SPACE(10); XB[P]:=READ;    HAR 2900
                     PRINT(XB[P],3,2); SPACE(10); BRB[P]:=READ; PRINT(BRB[P],3,2); HAR 3000
                     SPACE(9); BRB[P]:= BRB[P]* GX2; PRINT(BRB[P],3,4);    HAR 3100
                     XB[P]:= XB[P]* LMB;                         'END';     HAR 3200
       PAPERTHROW ;                                                        HAR 3300
              POLYFIT(XA,BRA,MA,TA,AVA,SCNA) ;     POLYFIT(XB,BRB,MB,TB,AVB,SCNB) ; HAR 3400
              A:=READ; B:=READ; A:=A*0.0254; B:=B*0.0254;                  HAR 3500
              'COMMENT' B IS OUTER ROTOR SURFACE RADIUS . ALL RADII IN INCHES ; HAR 3600
              AC:= 0.01745329252 ;                                         HAR 3700
              'FOR' P:= 1 'STEP' 1 'UNTIL' NA 'DO'                         HAR 3800
              'BEGIN' AM[P]:=READ; AN[P]:= AC*AM[P] ;           'END';     HAR 3900
       'FOR' R:= B,A 'DO'                                                  HAR 4000
       'BEGIN'   'FOR' P:= 1 'STEP' 1 'UNTIL' NA 'DO'                      HAR 4100
          BT[P]:= BR[P]:= BX[P]:= PHI[P]:= PHIP[P]:=0.0;                   HAR 4200
          B3:= B/A; B4:=B5:=B7:= B3*B3; B6:=B4*B4; BA:= B/A;              HAR 4300
          RA:= R/A; RA1:= RA*RA; RB:= B/R; RB1:= RB*RB;                   HAR 4400
          Q:= -1 ;      PVI:= 0.0 ;                                        HAR 4500
       'FOR' K:= 1 'STEP' 1 'UNTIL' TIMES 'DO'                            HAR 4600
       'BEGIN' 'INTEGER' N ;   N:=READ ;                                   HAR 4700
          AGAIN : Q:= Q+2 ;                                                HAR 4800
          COEFSUM(Q,1,TA,AVA,XA,BRA,SCNA,COEFA) ;                         HAR 4900
          COEFSUM(Q,1,TB,AVB,XB,BRB,SCNB,COEFB) ;                         HAR 5000
          COEFB[1,1]:=COEFA[1,1] + COEFB[1,1]; COEFB[2,1]:=COEFA[2,1] + COEFB[2,1]; HAR 5100
          COEFA[1,1]:=COEFA[1,1]/B;    COEFA[2,1]:=COEFA[2,1]/B;          HAR 5200
          COEFB[1,1]:=COEFB[1,1]/A;    COEFB[2,1]:=COEFB[2,1]/A;          HAR 5300
```

```
        ML:= 1.0/(B4 - 1);   B4:= B6*B4;    ML:= (A/R) * ML;                HAR 5400
        PN1:= COEFB[2,1] - B7 * COEFA[2,1] ;                                HAR 5500
        QN1:= COEFB[1,1] - B7 * COEFA[1,1] ;                                HAR 5600
        B7:= B5 * B7 ;                                                      HAR 5700
        PN2:= B3 * COEFB[2,1] - BA * COEFA[2,1] ;                           HAR 5800
        QN2:= B3 * COEFB[1,1] - BA * COEFA[1,1] ;                           HAR 5900
        B3:= B5 * B3 ;                                                      HAR 6000
        PN1:= RA * PN1 ;  QN1:= RA * QN1 ;    RA:= RA1 * RA ;               HAR 6100
        PN2:= RB * PN2 ;  QN2:= RB * QN2 ;    RB:= RB1 * RB ;               HAR 6200
        PN:= ML * (PN1 + PN2) ;      QN:= -ML * (QN1 + QN2) ;               HAR 6300
        RN:=-ML * (PN1 - PN2) ;     SN:= -ML * (QN1 - QN2) ;               HAR 6400
        PHI1:= PN * (R/Q) ;      PHI2:= QN * (-R/Q) ;                       HAR 6500
        'FOR' P:= 1 'STEP' 1 'UNTIL' NA 'DO'                                HAR 6600
        'BEGIN' RI:= Q * AN[P];    CC[P]:= COS(RI) ;    CS[P]:= SIN(RI) ;   HAR 6700
        BT[P]:= BT[P] + PN * CS[P] + QN * CC[P] ;                          HAR 6800
        BR[P]:= BR[P] + RN * CC[P] + SN * CS[P] ;                          HAR 6900
        PHI[P]:= PHI[P] + PHI1 * CC[P] + PHI2 * CS[P] ;                    HAR 7000
        POLAR (BT[P],BR[P],BX[P],PA[P] );                                  HAR 7100
        'END' ;                                                            HAR 7200
        PVI:= PVI + PN * SN + QN * RN ;                                    HAR 7300
        'IF' Q = 2*N - 1 'THEN' 'GOTO' OUT 'ELSE' 'GOTO' AGAIN ;          HAR 7400
OUT:PAPERTHROW ;                                                            HAR 7500
        WRITETEXT('('''('1C')' RESULTS%FROM%CALCULATION%OF%FLUX%DENSITY%AND%POWER%   HAR 7600
        FLOW%-%FIELD%THEORY%APPROACH '('2C')' THIN%ROTOR '('25S')' SURFACE    HAR 7700
        '('30S')' EXCITATION%=%'('2C')' RADIUS%FOR%CALCULATIONS,R%(M)%= ')') ;  HAR 7800
        PRINT(R,2,7);  WRITETEXT('('''('2C')'%ANGLE'('6S')' RAD.FLUX%DENSITY   HAR 7900
        '('4S')' TAN.FLUX%DENSITY'('8S')' RESULTANT'('8S')' PHASE%ANGLE '('3S')'   HAR 8000
        MAG.POTENTIAL- '('3S')' POYNTING%VECTOR '('1C')'% (DEG) '('10S')' (TESLA)   HAR 8100
        '('13S')' (TESLA) '('14S')' (TESLA) '('10S')' (DEGREES) '('4S')' AT%RAD.R%   HAR 8200
        (AMP) '('6S')' -SHAPE%OF '('2C')'')') ;                            HAR 8300
        'FOR' P:= 1 'STEP' 1 'UNTIL' NA 'DO'                               HAR 8400
        'BEGIN' NEWLINE(1); PRINT(AM[P],3,1); SPACE(7); PRINT(BR[P]*Y,0,4);   HAR 8500
        SPACE(7); PRINT(BT[P]*Y,0,4); SPACE(7); PRINT(BX[P]*Y,0,4); SPACE(7);   HAR 8600
        PRINT(PA[P],3,2); SPACE(5); PRINT(PHI[P]*MUI*Y,0,4); SPACE(4);     HAR 8700
        PRINT(BR[P] *BT[P],0,4);                                           HAR 8800
        'END' ;                                                            HAR 8900
        WRITETEXT('('''('2C')'POWER%FLOW(IN%SYN.%WATTS) = '('2S')'')');    HAR 9000
        Z:= PVI * R * R * (250.0/(2.54*OM)); PRINT(Z,0,6);                 HAR 9100
        WRITETEXT('('''('2C')' TORQUE%EQUIVALENT%TO%POWER%FLOW%(N-M)%=%')');   HAR 9200
        PRINT(( Z/OM ),4,7);                                               HAR 9300
        WRITETEXT('('''('2C')'ORDER%OF%HARMONIC%='('1S')'')'); PRINT(Q,3,0);   HAR 9400
        WRITETEXT('('''('10S')'NUMBER%OF%HARMONICS%='('1S')'')'); PRINT(N,3,0);   HAR 9500
    'END' ;                                                                HAR 9600
  'END' ;                                                                  HAR 9700
'END' ;                                                                    HAR 9800
;                                                                          HAR 9900
```

3

Analogue and mathematical models

Teledeltos Plots

In the experimental machine the profile of poles plays a crucial role in the flux distribution in general and, more importantly, the leakage and fringing at the pole tips. The classical means of the use of Teledeltos (paper) plots are ideally suited to study the above effects in a two-dimensional field distribution.

Plots Without Rotor

The air region enclosed by the magnetic circuit of the machine *without* a rotor was considered first. Symmetry was assumed and plots were obtained for one-quarter of the field system only, using a configuration shown in Fig.AP3.1[22].

The values of the conjugate potential ψ_θ, along the pole surface A,B,C and a point D on the side are as shown, where the potential at D is the same as at X; the contour of the pole providing flux/pole that crosses the interpolar axis accordingly consists of the portion A,B,C,D, were measured at equal intervals.

Note: the paper model profile/boundary covered the area ABCDMN X X′O

Fig.AP3.1 : Teledeltos plot model for one-quarter section of the field system

---

[22]In effect, a *conjugate* system of field plotting was employed in which the iron surfaces were treated as flow line boundaries and the field winding was replaced by a current kernel. The position of the kernel (point K in the figure) was determined by the extrapolation of the iron filing plots (described in Appendix IV, Fig.AP4.1b) in the interpolar region into the winding space.

A graph of ψ_θ and $\partial\psi_\theta/\partial\theta$ along ABCD is given in Fig.AP3.2 which shows that the B_r distribution, being equivalent in magnitude to $\partial\psi_\theta/\partial\theta$ in the present case, under the pole surface or the airgap of the machine invariably contains a peak at the pole tip C where the boundary profile changes abruptly[23].

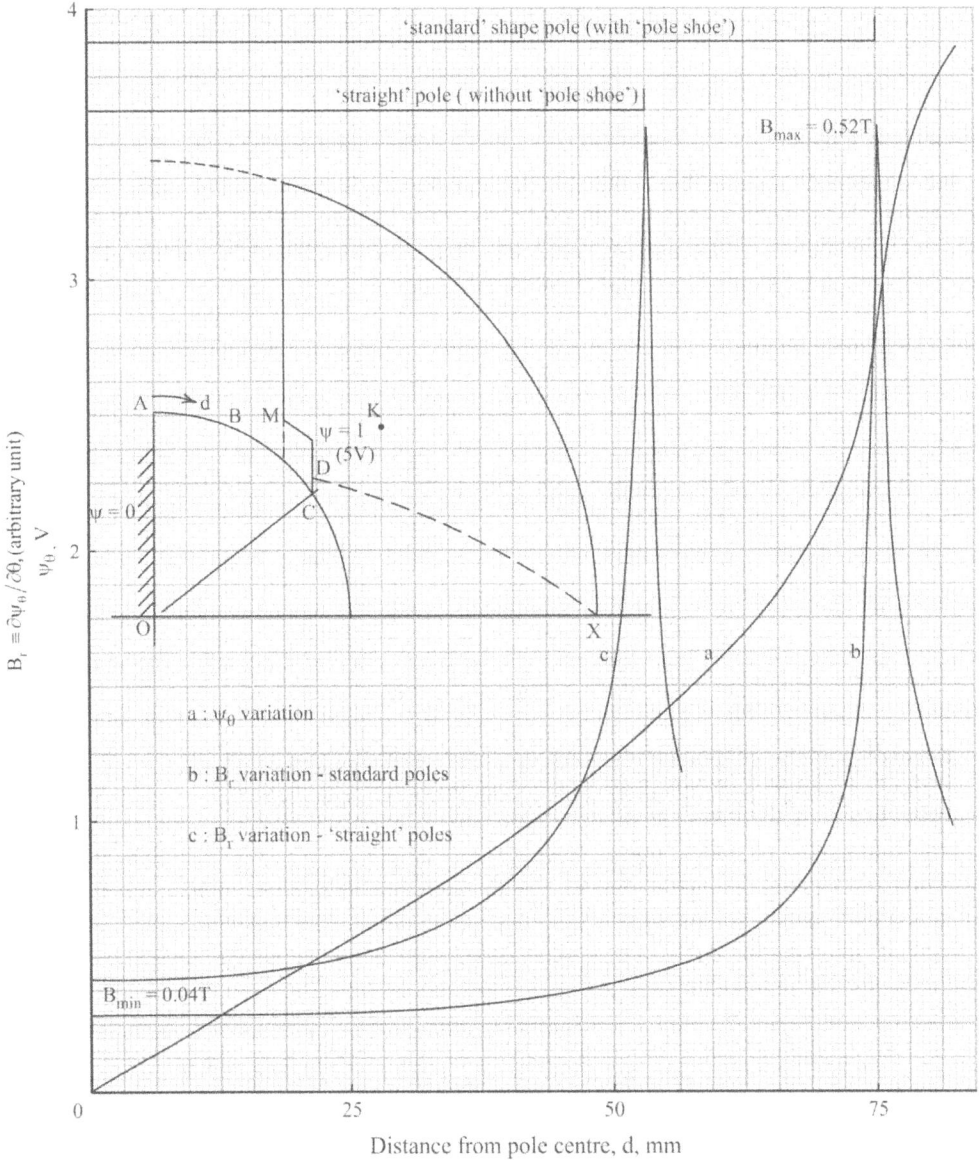

Fig.AP3.2 : Plots of ψ_θ and $\partial\psi_\theta/\partial\theta$ vs. distance from pole centre

---

[23]This corresponds nearly to the point of inflexion in the plot of Fig.AP3.2.

With the knowledge of the potential at X, Fig.AP3.1 or inset in Fig.AP3.2, the B_r distribution was computed for an mmf of 2,200 A/pole (corresponding to an excitation of 1.0 A in the experimental machine). The ratio of peak value to that under the pole centre compares favourably with the measured values in the machine obtained using a calibrated Hall probe[24].

Plots on the same scale were also obtained for 'straight' poles (contour ABM'MN in Fig.AP3.1), disregarding the 'pole shoe', and the B_r variation for this is also given in Fig.AP3.2. It is seen that the peak still occurs at the pole tip, M', but the ratio of maximum to minimum flux density is reduced to 8.7 as against 13.0 with the 'standard'- shaped poles as in the actual field system.

Analytical Field Plots

The above studies were supplemented by analysing a purely theoretical model of the region within the poles to determine the field distribution more accurately and to some extent overcome the errors and limitations of analogue plots[50, 51][25].

Both iron-filing and Teledeltos plots revealed that the arc formed by extending the circle containing the pole surfaces into the interpolar region, is a flow-line boundary in the absence of the rotor. This formed the basis for the analytical model which comprised a circular region.

Conformal transformation[26]

The field within a circle of radius R_o in z-plane, in which the pole arcs are represented by two 'electrodes' on the perimeter as depicted in Fig.AP3.3(a) would be transformed to a uniform field in a χ-plane (Fig.AP3.3(c)) by the transformation

$$z^* = \frac{R_o}{j}\left(\frac{1+j\sqrt{k}\ \text{sn}\ \chi}{1-j\sqrt{k}\text{sn}\ \chi}\right)^{27} \tag{AP3.1}$$

where   z is any point within the field of z-plane, $z = r\,\varepsilon^{j\theta}$

and   sn χ is a Jacobian elliptic function to modulus k $(=\tan^2 \alpha/2)$,

α being one-half the pole arc as shown in Fig.AP3.3(a)

---

[24]The calculations revealed that, in addition to the B_r waveform at the pole tip, the pole or yoke geometry results in about one-third flux leaking away owing to low permeability of the region.

[25]The analytical field plots also served the important requirement of checking the accuracy f the computer program developed for the computation of B_θ from the B_r waveforms.

[26]See, for example

S.C.Bhargava and M.J.Jevons: Mapping of magnetic field due to DC excited unsaturated salient-pole field system, J. Inst. Engrs (I), Vol. 64, pt. EL 4, 1984, pp 181-85.

[27]The usual transformation from the z-plane through w-plane to the χ-plane, using elliptic functions, gives electrodes parallel to the ψ-plane. In the present case, the anti-clockwise rotation by $\pi/2$ is obtained by using z^* (conjugate function of z) in eqn.(AP3.1).

This is done to have electrodes in the χ-plane to suit appropriately the situation in the z-plane: the potential $\psi = 0$ in the χ-plane is also equivalent to $\psi = 0$ on one of the electrodes in the z-plane.

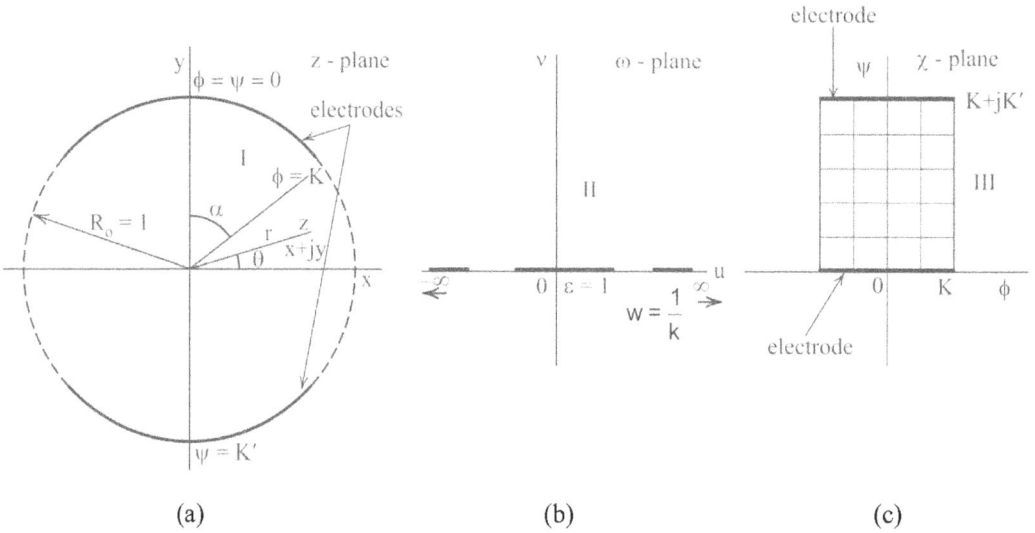

Fig.AP3.3 : Representation of pole arcs of the field system as electrodes

Calculation of field components

The magnetic excitation $\overline{H}$ in the region is given by

$$\overline{H} = - \text{ grad } \psi$$

that is

$$H_r = - \frac{\partial \psi}{\partial r} \quad \text{and} \quad H_\theta = - \frac{1}{r}\frac{\partial \psi}{\partial \theta} = - \frac{\partial \phi}{\partial r}^{28}$$

Hence for a two dimensional $\overline{H}$ field, using complex variable notation

$$\overline{H} = H_r + j\,H_\theta = -\frac{\partial}{\partial r}(\psi + j\phi) = j\frac{\partial}{\partial r}(\phi + j\psi)$$

or

$$\overline{H} = j\frac{\partial \chi}{\partial r} \tag{AP3.2}$$

Now for any point $z = r\,\varepsilon^{j\theta}$, eqn.(AP3.1) can be written

$$j\,r\,\varepsilon^{-j\theta} = R_o \left\{ \frac{1 + j\sqrt{k}\ \text{sn } \chi}{1 - j\sqrt{k}\ \text{sn } \chi} \right\} \tag{AP3.3}$$

---

[28]In cylindrical coordinates and considering the χ-plane

$$x = \phi + j\,\psi$$

therefore

$$\frac{\partial \phi}{\partial r} = \frac{1}{r}\frac{\partial \psi}{\partial \theta}$$

and

$$\frac{\partial \psi}{\partial r} = -\frac{1}{r}\frac{\partial \phi}{\partial \theta}$$

Differentiating with respect to r

$$j\, r\, \varepsilon^{-j\theta} = \frac{j\, 2R_o \sqrt{k}\ sn'\, \chi}{(1 - j\sqrt{k}\ sn\ \chi)^2}\, \frac{\partial \chi}{\partial r} \qquad\qquad \text{(AP3.4)}$$

or

$$\varepsilon^{-j\theta} = \frac{2R_o \sqrt{k}\ sn'\, \chi}{(1 - j\sqrt{k}\ sn\ \chi)^2}\left(\frac{\partial \phi}{\partial r} + j\frac{\partial \psi}{\partial r}\right) \qquad\qquad \text{(AP3.5)}$$

since

$$\frac{\partial}{\partial r} = \frac{\partial}{\partial \chi}\frac{\partial \chi}{\partial r} = \frac{\partial}{\partial \chi}\left(\frac{\partial \phi}{\partial r} + j\frac{\partial \psi}{\partial r}\right)$$

The solution of eqn.(AP3.5), giving values of $\dfrac{\partial \phi}{\partial r}$ and $\dfrac{\partial \psi}{\partial r}$, then provides the required field components in accordance with the relationships derived in eqn.(AP3.2), in terms of the known quantities R_o, r and θ. This is done as follows:

Let

$$\frac{2R_o \sqrt{k}\, sn'\, \chi}{(1 - j\sqrt{k}\ sn\ \chi)^2} = C + jD \qquad \text{and} \qquad \frac{\partial \phi}{\partial r} + j\frac{\partial \psi}{\partial r} = A + jB$$

Then, from eqn.(AP3.5),

$$\cos\theta - j\sin\theta = (C + j\, D)\, (A + j\, B)$$
$$= CA + j\, BC + j\, DA - DB$$

or in matrix form, equating real and imaginary parts

$$\begin{bmatrix} \cos\theta \\ -\sin\theta \end{bmatrix} = \begin{bmatrix} C & -D \\ D & C \end{bmatrix}\begin{bmatrix} A \\ B \end{bmatrix}$$

Now let $\Delta = C^2 + D^2$ giving

$$A = (C\cos\theta - D\sin\theta)/\Delta$$
$$B = (-C\sin\theta - D\cos\theta)/\Delta$$

or

$$H_r = -\frac{\partial \psi}{\partial r} = -B$$

and

$$H_\theta = -\frac{\partial \phi}{\partial r} = -A$$

The value of $sn\ \chi$ is obtained using eqn.(AP3.3) and $sn'\, \chi$ calculated from

$$sn'\chi = cn\ \chi\ dn\ \chi$$

where

$$cn\ \chi = \sqrt{1 - sn^2\chi}$$

and

$$dn = \sqrt{(1 - k^2 sn^2\chi)}$$

Calculations were carried out[29] for values of H_r, H_θ and H for different angles and radii and the variation of the field components for r = 0.25 to r = 0.99 (almost representing the pole surface) is shown in Figs.AP3.4, AP3.5 and AP3.6 which

---

[29]The procedures to calculate the values of $sn\ \chi$ etc. to modulus k and k' were based on the information available in [52] and [53].

demonstrate the effect of pole tip on the field components and resultant field, H. A field plot eliciting the pattern of flow lines in one-quarter region (z-plane) is given in Fig.AP3.7.

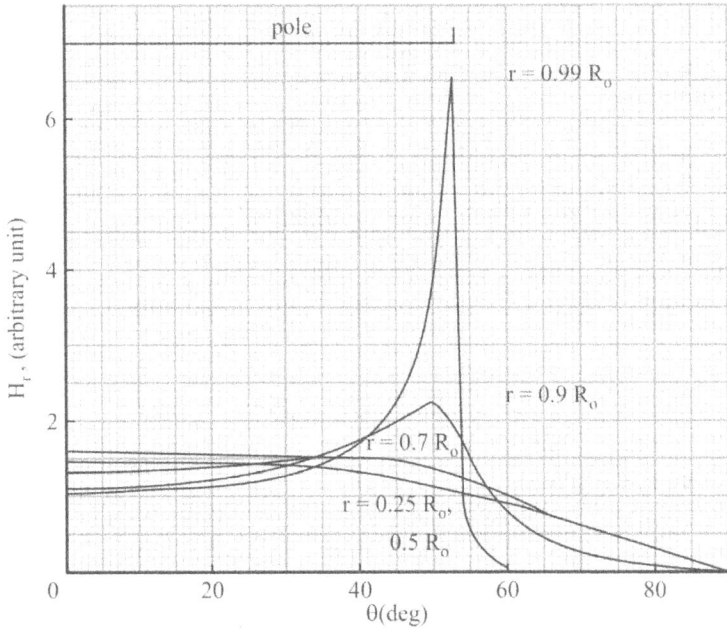

Fig.AP3.4 : Variation of radial component, H_r

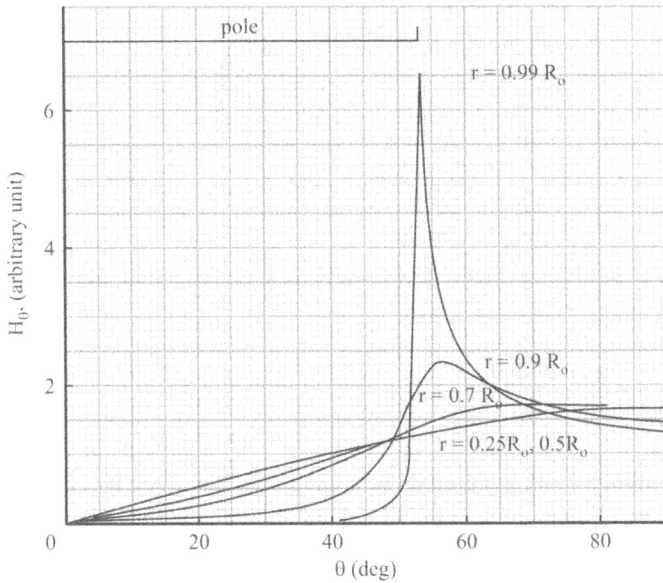

Fig.AP3.5 : Variation of peripheral component, H_θ

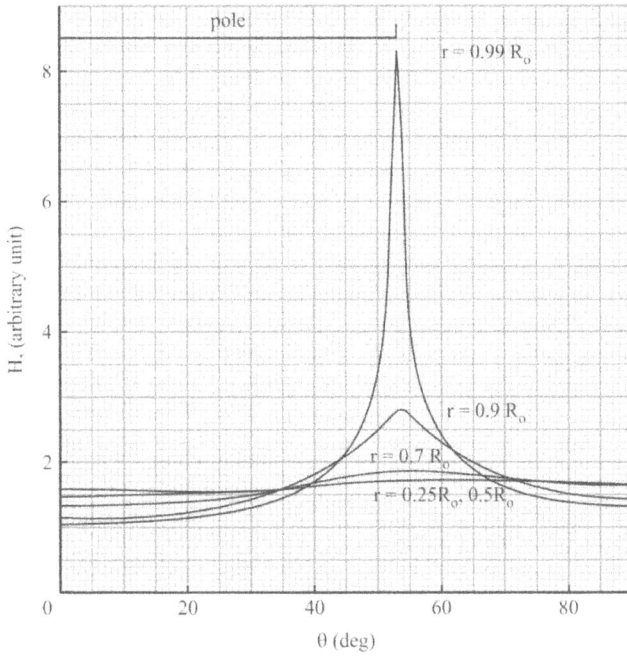

Fig.AP3.6 : Variation of resultant, H

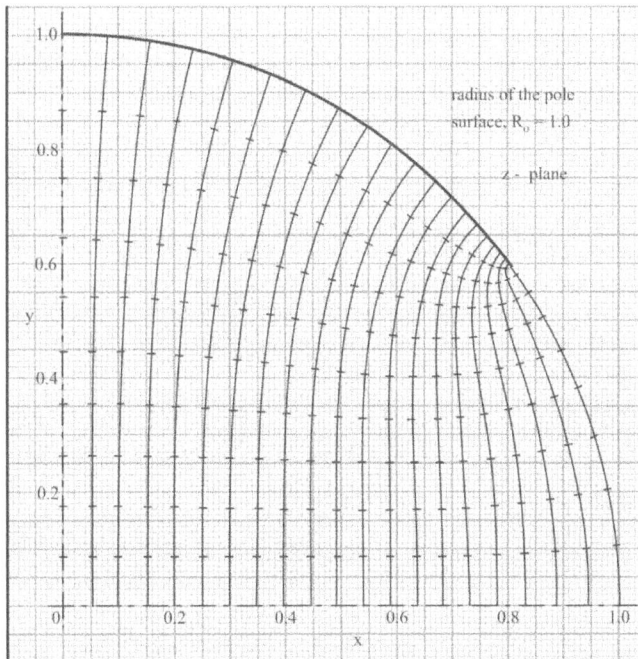

Fig.AP3.7 : Field plot showing pattern of flow lines

4

Iron filling patterns

The simple technique of using iron filings formed the basis of *visually* observing leakage flux distribution in the interpolar and arbor regions of the experimental machine.

The Process

The patterns were obtained by sprinkling iron filings on a piece of DUOSTAT paper, cut to the shape of the region of interest and positioned in the machine, the latter having been located in a suitable "dark room" for carrying out the tests.

The field winding was excited with currents ranging from 0.05 A to 0.3 A when the iron filings acquired the shape of flux lines in the region. The pattern that was formed was exposed to light for a short time, the (photographic) paper removed gently and developed and "fixed" appropriately leaving a permanent record of the pattern under varying conditions.

The tests were performed first with no rotor present in between the poles and later with a vicalloy rotor[1].

Test Results

The patterns, obtained for various representative test conditions, are illustrated in Figs.AP4.1 through AP4.5. The pattern in Fig.AP4.1 corresponds to the field due to the *unsaturated* field system alone (yoke and poles), at a small excitation of 0.05 A. One half of this pattern and its diagrammatic representation is depicted in Fig.AP4.2.

---

[1]For convenience of handling, a vicalloy annulus of radial thickness of about 15 mm was used; this, however, did not jeopardize the qualitative effect of rotor hysteresis on flux distribution in the regions of interest.

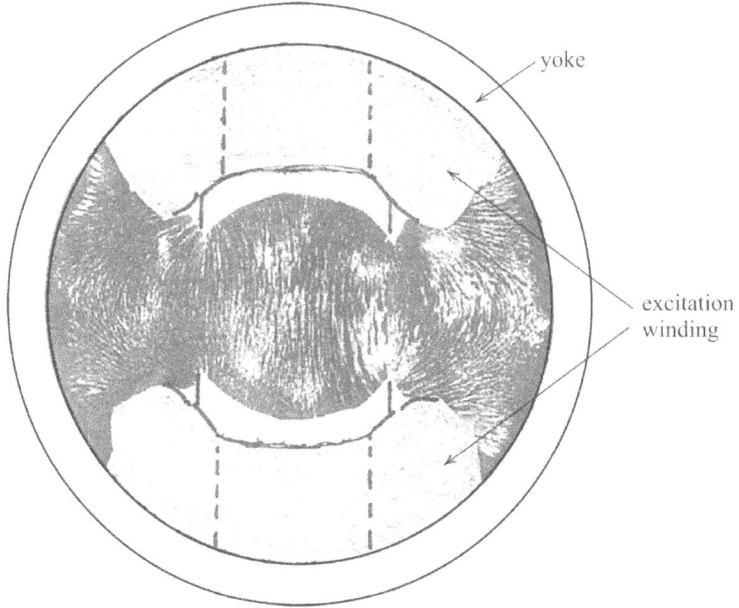

Fig.AP4.1 : Iron filing pattern of field system alone

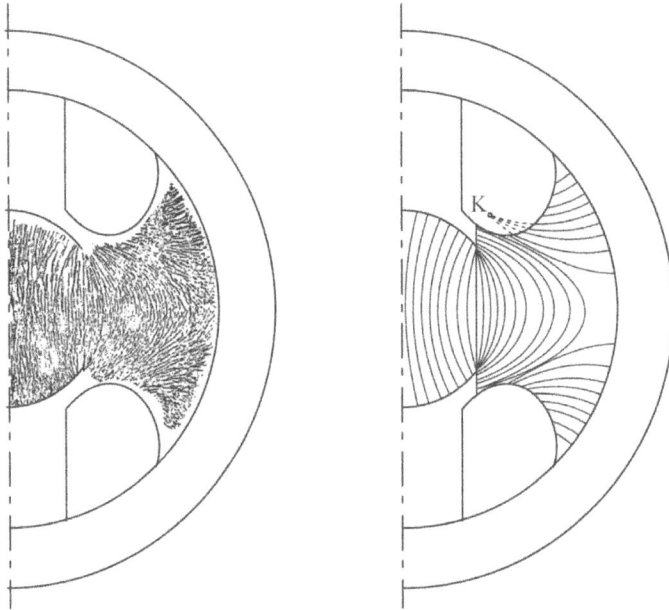

(a) excitation 0.05 A: no rotor

(b) diagrammatic representation

K : kernel

Fig.AP4.2 : Right hand part of iron filing pattern due to field system alone

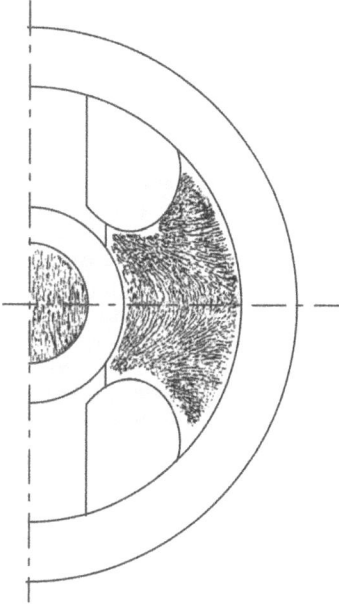

Fig.AP4.3 : Iron filing pattern of leakage flux of
            vicalloy rotor

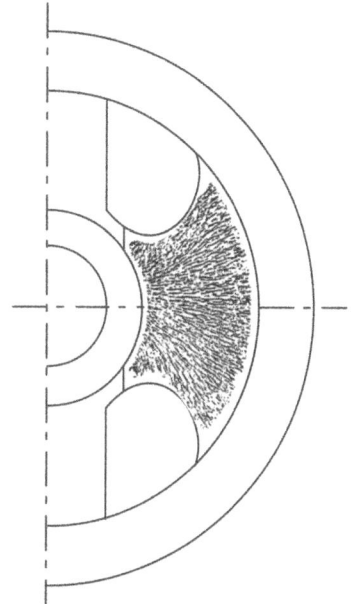

Fig.AP4.4 : Iron filing pattern of residual flux
            of vicalloy rotor magnetisation

excitation 0.3 A

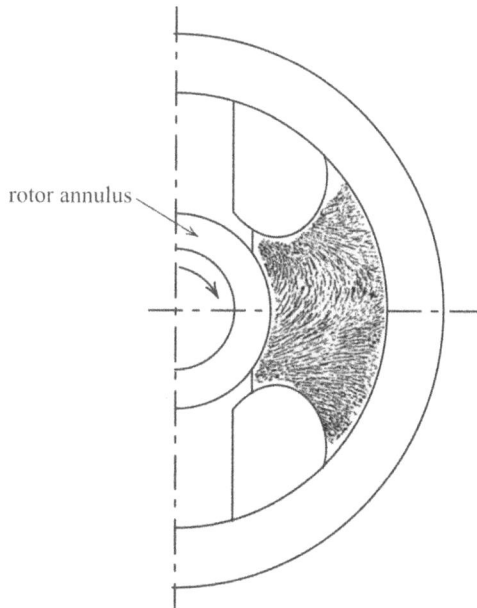

rotor annulus

Fig.AP4.5 : Iron filing pattern due to vicalloy rotor showing spatial hysteresis at 0.3 A

The pattern of Fig.AP4.3 indicates the 'independent' rotor magnetisation in the form of residual magnetism when the excitation is switched off and is seen to be in space quadrature with the field due to field system alone shown in Fig.AP4.1.

Spatial hysteresis

The spatial hysteresis effect, exhibited by the pattern in Fig.AP4.5, was produced by rotating the rotor annulus by hand some 4/5 times in a clockwise direction and holding it stationary without it slipping back. Observe that the flux has shifted towards the lagging pole tip and the leakage in the direction of the yoke is increased. The effect was observed to be most pronounced at the excitation of 0.3 A and to diminish to the case of non-hysteretic pattern at about 1.0 A.

5

Field theory calculations

Evaluation of Coefficients for B_θ Waveforms Derived from B_r Waveforms

Using Field Theory

In the airgap or arbor region of the machine, $\overline{H}$ is irrotational (assuming a non-time dependent field) and can be expressed as the negative gradient of a magnetic scalar potential, governed by Laplace equation.

Let the general solution of the Laplace equation in the region, based on separation of variables, be

$$\phi = \sum_{n=1}^{\infty} \left(A_n\ r^n + B_n\ r^{-n} \right)\cos n\theta + \left(D_n\ r^n + E_n\ r^{-n} \right)\sin\ n\theta \qquad \text{(AP5.1)}$$

Then $\qquad H_r = -\dfrac{\partial\phi}{\partial r}$

$$= -\sum_{n=1}^{\infty} n\left(A_n\ r^{n-1} - B_n\ r^{-n-1} \right)\cos n\theta - n\left(D_n\ r^{n-1} - E_n\ r^{-n-1} \right)\sin\theta \quad \text{(AP5.2)}$$

$$B_r = \mu_o\ H_r$$

and $\qquad H_\theta = -\dfrac{1}{r}\dfrac{\partial\phi}{\partial\theta}$

$$= \sum_{n=1}^{\infty} n\left(A_n\ r^{n-1} + B_n\ r^{-n-1} \right)\sin n\theta - n\left(D_n r^{n-1} + E_n\ r^{-n-1} \right)\cos n\theta \qquad \text{(AP5.3)}$$

$$B_\theta = \mu_o\ H_\theta$$

From the harmonic analysis of measured B_r waveforms, at $r = a$

$$B_{r_a} = A_n'\ \sin n\theta + B_n'\ \cos n\theta \qquad\qquad \text{(AP5.4a)}$$

and at $r = b$

$$B_{r_b} = A_n''\ \sin n\theta + B_n''\ \cos n\theta \qquad\qquad \text{(AP5.4b)}$$

where, if $a > b$, B_{r_a} is obtained by adding to the expansion of B_{r_b} the expansion of ΔB_r waveform.

The coefficients A_n', . . . , B_n'' are therefore known.

Using eqn.(AP5.2) at $2\,r = a$ and b,

$$B_{r_a} = -\sum_{n=1}^{\infty} \mu_o \left[n\left(A_n\,a^{n-1} - B_n\,a^{-n-1}\right)\cos n\theta - n\left(D_n\,a^{n-1} - E_n\,a^{-n-1}\right)\sin n\theta \right] \quad \text{(AP5.5a)}$$

$$B_{r_b} = -\sum_{n=1}^{\infty} \mu_o \left[n\left(A_n\,b^{n-1} - B_n\,b^{-n-1}\right)\cos n\theta - n\left(D_n\,b^{n-1} - E_n\,b^{-n-1}\right)\sin n\theta \right] \quad \text{(AP5.5b)}$$

Equating the expressions on the right hand side of eqns.(AP5.4a) and (AP5.4b) to eqns.(AP5.5a) and (AP5.5b), respectively, and solving for A_n, B_n, D_n and E_n

$$A_n = \frac{a^{-n+1}}{n\,\mu_o \left[(b/a)^{2n} - 1 \right]} \left[B_n' - (b/a)^{n+1}\,B_n'' \right] \quad \text{(AP5.6a)}$$

$$B_n = \frac{a^{n+1}}{n\,\mu_o \left[(b/a)^{2n} - 1 \right]} \left[(b/a)^{2n}\,B_n' - (b/a)^{n+1}\,B_n'' \right] \quad \text{(AP5.6b)}$$

$$D_n = \frac{a^{-n+1}}{n\,\mu_o \left[(b/a)^{2n} - 1 \right]} \left[A_n' - (b/a)^{n+1}\,A_n'' \right] \quad \text{(AP5.6c)}$$

$$E_n = \frac{a^{n+1}}{n\,\mu_o \left[(b/a)^{2n} - 1 \right]} \left[(b/a)^{2n}\,A_n' - (b/a)^{n+1}\,A_n'' \right] \quad \text{(AP5.6d)}$$

At any radius $r = R$ in the region, B_θ from eqn.(AP5.3) is given by

$$B_\theta = \sum_{n=1}^{\infty} \mu_o \left[n\left(A_n\,R^{n-1} + B_n\,R^{-n-1}\right)\sin n\theta - n\left(D_n\,R^{n-1} + E_n\,R^{-n-1}\right)\cos n\theta \right]$$

$$\text{(AP5.7)}$$

with A_n, . . . E_n evaluated in eqns.(AP5.6).

Let $B_\theta = P_n \sin n\theta + Q_n \cos n\theta$ $\qquad\qquad$ (AP5.7a)

where $P_n = \mu_o\,n\left(A_n\,R^{n-1} + B_n\,R^{-n-1}\right)$ $\qquad\qquad$ (AP5.8a)

and $Q_n = -\mu_o\,n\left(D_n\,R^{n-1} + E_n\,R^{-n-1}\right)$ $\qquad\qquad$ (AP5.8b)

Then substituting for A_n, . . . E_n, in expressions for P_n and Q_n

$$P_n = \frac{1}{\left[(b/a)^{2n} - 1 \right]} \left\{ (R/a)^{n-1}\left[B_n' - (b/a)^{n+1}B_n'' \right] + (a/R)^{n+1}\left[(b/a)^{2n}B_n' - (b/a)^{n+1}B_n'' \right] \right\}$$

$$\text{(AP5.9a)}$$

$$Q_n = \frac{-1}{\left[(b/a)^{2n} - 1 \right]} \left\{ (R/a)^{n-1}\left[A_n' - (b/a)^{n+1}A_n'' \right] + (a/R)^{n+1}\left[(b/a)^{2n}A_n' - (b/a)^{n+1}A_n'' \right] \right\}$$

$$\text{(AP5.9b)}$$

P_n and Q_n may be further written

$$P_n = ML\left\{P_{n_1} + P_{n_2}\right\} \tag{AP5.9c}$$

$$Q_n = -ML\left\{Q_{n_1} + Q_{n_2}\right\} \tag{AP5.9d}$$

where

$$ML = \frac{1}{\left[(b/a)^{2n} - 1\right]} \tag{AP5.10a}$$

$$P_{n_1} = (R/a)^{n-1}\left[B_n' - (b/a)^{n+1}B_n''\right] \tag{AP5.10b}$$

$$P_{n_2} = (a/R)^{n+1}\left[(b/a)^{2n}B_n' - (b/a)^{n+1}B_n''\right] \tag{AP5.10c}$$

$$Q_{n_1} = (R/a)^{n-1}\left[A_n' - (b/a)^{n+1}A_n''\right] \tag{AP5.10d}$$

$$Q_{n_2} = (a/R)^{n+1}\left[(b/a)^{2n}A_n' - (b/a)^{n+1}A_n''\right] \tag{AP5.10e}$$

Eqn.(AP5.7a) thus reduces to

$$B_\theta = ML\left\{\left[P_{n_1} + P_{n_2}\right]\sin n\theta - \left[Q_{n_1} + Q_{n_2}\right]\cos n\theta\right\}^{31} \quad \text{(AP5.11)}$$

Calculation of Magnetic Scalar Potential

The assumed expression for the scalar potential in the region is, using eqn.(AP5.1)

$$\phi = (A_n r^n + B_n r^{-n})\cos n\theta + (D_n r^n + E_n r^{-n})\sin n\theta$$

where A_n, . . E_n have been derived previously in terms of A_n', ... B_n'' following harmonic analysis of B_r waveforms at radii a and b.

Accordingly, the magnetic scalar potential variation can be deduced at the two redii.

---

[31]These slight modifications are directed to writing of a computer program later.

6

Poynting vector and other forms

Poynting Vector

Derivation

Using the identity

$$\overline{\nabla} \cdot (\overline{E} \times \overline{H}) = \overline{H} \cdot (\overline{\nabla} \times \overline{E}) - \overline{E} \cdot (\overline{\nabla} \times \overline{H}) \qquad \text{(AP6.1)}$$

Now

$$\overline{\nabla} \times \overline{E} = -\frac{\partial \overline{B}}{\partial t}$$

and

$$\overline{\nabla} \times \overline{H} = \overline{J} + \frac{\partial \overline{D}}{\partial t} = \rho \overline{E} + \frac{\partial \overline{D}}{\partial t} \qquad \text{(AP6.2)}$$

Therefore

$$\overline{\nabla} \cdot (\overline{E} \times \overline{H}) = -\overline{H} \cdot \frac{\partial \overline{B}}{\partial t} - \overline{E} \cdot \frac{\partial \overline{D}}{\partial t} - \rho \overline{E} \cdot \overline{E} \qquad \text{(AP6.3)}$$

When μ and ε are constant

$$\overline{H} \cdot \frac{\partial \overline{B}}{\partial t} = \mu \overline{H} \frac{\partial \overline{H}}{\partial t} = \frac{\mu}{2} \frac{\partial (\overline{H} \cdot \overline{H})}{\partial t} = \frac{\mu}{2} \frac{\partial H^2}{\partial t} \qquad \text{(AP6.4)}$$

and similarly

$$\overline{E} \cdot \frac{\partial \overline{D}}{\partial t} = \frac{\varepsilon}{2} \frac{\partial E^2}{\partial t} \qquad \text{(AP6.5)}$$

Hence

$$\overline{\nabla} \cdot (\overline{E} \times \overline{H}) = -\frac{\partial}{\partial t}\left(\frac{\mu}{2} H^2 + \frac{\varepsilon}{2} E^2\right) - \rho E^2 \qquad \text{(AP6.6)}$$

Associated with the time derivative, $\partial/\partial t$, the two terms within the brackets represent the *density* of energy stored in the magnetic and electric fields, respectively, whilst the term $-\rho E^2$ is equivalent to ohmic power loss per unit volume or Joulean heating as a result of flow of conduction current, ρE.

The vector $(\overline{E} \times \overline{H})$, representing energy flow in a given volume, is called **Poynting vector**, usually denoted by $\overline{S}$.

Sillar's Vector

In a 'very slowly' varying field without current source

$$\overline{\nabla} \times \overline{H} = \overline{J} + \frac{\partial \overline{D}}{\partial t} = 0 \tag{AP6.7}$$

since $J = 0$ and displacement currents, D, are also zero.
Therefore

$$\overline{H} = -\overline{\nabla}\phi, \quad \phi \text{ being a scalar potential}$$

Then, from Poynting vector

$$\begin{aligned}
\overline{S} &= \overline{E} \times \overline{H} = \overline{E} \times (-\overline{\nabla}\phi) \\
&= -\overline{E} \times (\overline{\nabla}\phi) = (\overline{\nabla}\phi) \times \overline{E} \\
&= \overline{\nabla} \times (\phi\overline{E}) - \phi(\overline{\nabla} \times \overline{E})
\end{aligned} \tag{AP6.8}$$

from the identity $\overline{\nabla} \times (\phi\overline{E}) = (\overline{\nabla}\phi) \times \overline{E} + \phi(\overline{\nabla} \times \overline{E})$

Or

$$\overline{E} \times \overline{H} = \overline{\nabla} \times (\phi\overline{E}) - \phi\left(-\frac{\partial \overline{B}}{\partial t}\right) \qquad \text{since } \overline{\nabla} \times \overline{E} = -\frac{\partial \overline{B}}{\partial t}$$

or

$$\overline{S}' = \overline{E} \times \overline{H} - \overline{\nabla} \times (\phi\overline{E}) = \phi\frac{\partial \overline{B}}{\partial t} \tag{AP6.9}$$

and is called **Sillar's vector**.

Slepian Vector

Using Maxwell's equation

$$\overline{\nabla} \times \overline{E} = \frac{\partial \overline{B}}{\partial t}$$

Expressing $\overline{B}$ in terms of a magnetic vector potential, $\overline{A}$

$$\overline{\nabla} \times \overline{E} = -\frac{\partial(\overline{\nabla} \times \overline{A})}{\partial t}$$

Or

$$\overline{E} = -\frac{\partial \overline{E}}{\partial t}$$

Therefore

$$\overline{S} = \overline{E} \times \overline{H} = -\overline{\nabla}\phi \times \overline{H} - \frac{\partial \overline{A}}{\partial t} \times \overline{H}$$

$$= \phi\overline{\nabla} \times \overline{H} - \overline{\nabla} \times (\phi\overline{H}) - \frac{1}{2}\frac{\partial}{\partial t}(\overline{A} \times \overline{H})$$

Let

$$\overline{C} = \overline{\nabla} \times (\phi\overline{H}) \text{ and } \overline{\nabla} \times \overline{H} = \overline{J} + \frac{\partial \overline{D}}{\partial t}$$

Then

$$\overline{S}_1 = \overline{S} + \overline{C} = \phi\left(\overline{J} + \frac{\partial \overline{D}}{\partial t}\right) + \frac{1}{2}\frac{\partial}{\partial t}(\overline{H} \times \overline{A}) \tag{AP6.10}$$

and is called **Slepian vector**.

7

Derivation of torque expression(s) for hysteresis-reluctance motor

Reluctance Torque

To obtain the expression for torque for reluctance motor, the presence of the hysteresis ring is disregarded and the machine analysed as a simple reluctance motor. Later, the hysteresis ring would be assumed 'sitting' round the reluctance-motor rotor.

The cross-section of the motor is shown in Fig.AP7.1. With the hysteresis ring of radial thickness t considered, the actual mechanical airgap length would be δ. To derive reluctance torque, the airgap length at an angle θ would be $g(\theta)$ such that $g(\theta)$ on the reluctance-rotor (or direct) axis and over an arc 2α is given by

$$g(\theta)_{min} = t + \delta \qquad \text{(AP7.1)}$$

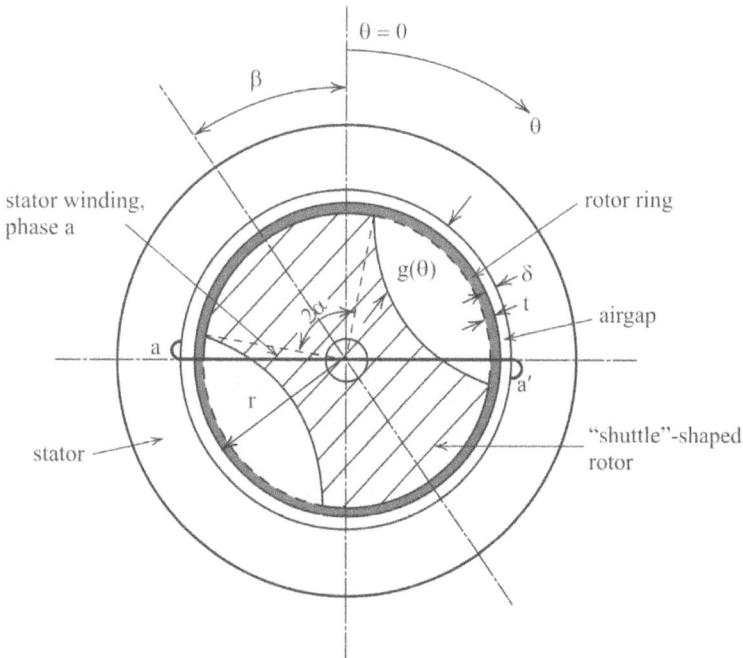

Fig.AP7.1 : Two-pole model of hysteresis-reluctance motor

As shown, (θ) is symmetrical about $\theta = -\beta$ and $\theta = -\beta + \pi/2$.

If the stator currents are balanced and the current in phase a is

$$I_a = I_m \sin \omega t \tag{AP7.2}$$

the magnetic potential across the airgap will be

$$F_\theta \frac{3 N_s I_m}{4} \sin (\omega t - \theta) \text{ A} \tag{AP7.3}$$

at any angle θ and for three-phase currents in sinusoidally distributed winding, with N_s number of turns in each stator winding.

The flux density in the airgap at angle θ is approximated by

$$B_\theta \frac{\mu_o}{g(\theta)} F_\theta \tag{AP7.4}$$

In eqn.(AP7.4), let $1/g(\theta)$ be expanded as Fourier series to account for the 'non-uniform' airgap.

Then, let

$$\frac{1}{g(\theta)} = h_o + h_2 \cos 2(\theta + \beta) + h_4 \cos 4(\theta + \beta) + \ldots \tag{AP7.5}$$

Now in actual operation, the rotor would rotate synchronously with an angular speed, ω rad/s, same as the stator rotating magnetic field. This suggests that β be expressed as a function of ω, and time t, plus a constant angle of lag between the two magnetic axes as shown in Fig.AP7.1. Therefore, let

$$-\beta = \omega t + \beta_o$$

or

$$\beta = -(\omega t + \beta_o) \tag{AP7.6}$$

Substituting for β in eqn.(AP7.5) and then $1/g(\theta)$ in eqn.(AP7.4),

$$B_\theta = \mu_o F_\theta \left[h_o + h_2 \cos 2(\theta - \omega t - \beta_o) + h_4 \cos 4(\theta - \omega t - \beta_o) + \ldots \right] \tag{AP7.7}$$

and substituting for F_θ from eqn.(AP7.3) in eqn.(AP7.7)

$$B_\theta = \frac{3 \mu_o N_s I_m}{4} \left[h_o \sin(\omega t - \theta) + h_2 \sin(\omega t - \theta)\cos 2(\theta - \omega t - \beta_o) \right.$$

$$\left. + h_4 \sin(\omega t - \theta) \cos 4(\theta - \omega t - \beta_o) + \ldots \right]$$

$$= \frac{3 \mu_o N_s I_m}{4} \left[h_o \sin(\omega t - \theta) + \frac{h_2}{2} \{\sin(3\omega t + 2\beta_o - 3\theta) - \sin(\omega t + 2\beta_o - \theta)\} \right.$$

$$\left. + \frac{h_4}{2} \{\sin(5\omega t + 4\beta_o - 5\theta) - \sin(3\omega t + 4\beta_o - 3\theta)\} + \ldots \right] \tag{AP7.8}$$

Flux linkage of stator winding a

The torque developed in the machine would be obtained as the rate of change of flux linkage in the airgap with respect to the angle β_o. To obtain the flux linkage of stator winding, a, first determine the flux linkage of a *single* turn at θ, 'returning' at $(\theta+\pi)$. Let r is the airgap radius and L is the axial length of the machine. Consider a small displacement $d\theta$. The flux, assumed to be radial in the airgap, links with an area r $d\theta$ L at the angle θ. For the single turn, the flux linkage would then be, between θ and $\pi+\theta$,

$$\lambda_{1_\theta} = \int_\pi^{\theta+\pi} B_\theta L\, r\, d\theta$$

$$= \int_\pi^{\theta+\pi} \frac{L\, r\, 3\mu_o\, N_s\, I_m}{4}\left[\theta h_o \sin(\omega t - \theta) + \frac{h_2}{2}\{\sin(3\omega t + 2\beta_o - 3\theta) - \sin(\omega t + 2\beta_o - \theta)\}\right.$$

$$\left. + \frac{h_4}{2}\{\sin(5\omega t + 4\beta_o - 5\theta) - \sin(3\omega t + 4\beta_o - 3\theta)\} + ...\right] d\theta$$

by substituting for B_θ from eqn.(AP7.8).

or

$$\lambda_{1_\theta} = \int_\pi^{\theta+\pi} \frac{L\, r\, 3\mu_o\, N_s\, I_m}{2}\left[-h_o \cos(\omega t - \theta) + \frac{h_2}{2}\right.$$

$$\left\{-\frac{1}{3}\cos(3\omega t + 2\beta_o - 3\theta) - \cos(\omega t + 2\beta_o - \theta)\right\}$$

$$\left. + \frac{h_4}{2}\left\{-\frac{1}{5}\cos(5\omega t + 4\beta_o - 5\theta) + \frac{1}{3}\cos(3\omega t + 4\beta_o - 3\theta)\right\} + ...\right] \qquad (AP7.9)$$

Now the number of phase a conductors in a band $d\theta$ is $(N_s \sin \theta)$,2) times $d\theta$. Thus, the total flux linage of *winding* a is

$$\lambda_{m_a} = \int_{-\pi}^0 \frac{N_s \sin\theta}{2}\lambda_{i_\theta}\, d\theta$$

$$= \frac{L\, r\, 3\mu_o\, N_s^2\, I_m}{4}\int_{-\pi}^0\left[-\frac{h_o}{2}\{\sin(2\theta - \omega t) + \sin\omega t\}\right.$$

$$-\frac{h_2}{12}\{\sin(3\theta - 3\omega t - 2\beta_o) + \sin(3\omega t - 2\theta + 2\beta_o)\}$$

$$+\frac{h_2}{4}\{\sin(2\theta - \omega t - 2\beta_o) + \sin(\omega t + 2\beta_o)\}$$

$$\left. + \;...\; \text{higher order terms}\right] d\theta$$

Neglecting higher-order harmonic terms, the flux linkage of winding a of *fundamental* frequency is

$$\lambda_{m_a} = \frac{L\,r\,3\,\mu_o\,N_s^2\,I_m}{4} \int_{-\pi}^{0} \left[-\frac{h_o}{2}\{\sin(2\theta - \omega t) + \sin\omega t\} \right.$$

$$\left. +\frac{h_2}{4}\{\sin(2\theta - \omega t - 2\beta_o) + \sin(\omega t + 2\beta_o)\} \right] d\theta$$

$$= K\,I_m \left[h_o\,\sin\omega t - \frac{h_2}{2}\sin\left(\omega t + 2\beta_o\right) \right] \qquad (AP7.10)$$

where

$$K = \frac{\pi\,L\,r\,3\,\mu_o\,N_s^2\,I_m}{8} \qquad (AP7.11)$$

The torque of the machine can be obtained using the expression

$$\tau = i\,\frac{d\lambda}{d\beta_o} \qquad (AP7.12)$$

where τ is the instantaneous torque, λ the total instantaneous linkage and i the instantaneous current. Let the stator linkage due to leakage flux of phase winding a alone be λ_{1_a} and independent of β_o. Then although the current in phase a, i_a, is time dependent, but independent of β_o, the torque due to linkage of phase a alone will be

$$\tau_a = i_a\,\frac{d\lambda_{m_a}}{d\beta_o} \qquad (AP7.13)$$

and total instantaneous torque due to the 3-phase winding would be

$$\tau = 3\,\tau_a \qquad (AP7.14)$$

From eqn.(AP7.10)

$$\frac{d\lambda_{m_a}}{d\beta_o} = -K\,I_m\,\frac{h_2}{2}\,2\cos\left(\omega t + 2\beta_o\right)$$

$$\therefore \qquad \tau_a = I_m\,\sin\omega t\,\left[-K\,I_m\,h_2\,\cos\left(\omega t + 2\beta_o\right) \right]$$

$$= -I_m^2\,\frac{K}{2}\,h_2\left[-\sin\left(2\beta_o\right) + \sin\left(2\omega t + 2\beta_o\right) \right]$$

and
$$\tau = 3\,\tau_a$$

$$= -\frac{3\,K\,h_2}{2}\,I_m^2\left[-\sin\left(2\beta_o\right) + \sin\left(2\omega t + 2\beta_o\right) \right] \qquad (AP7.15)$$

The average developed torque in phase winding a

$$T_a = \frac{1}{\pi}\int_{-\pi}^{0} \tau_a\,d(\omega t)$$

$$= \frac{1}{\pi}\left(-\frac{3\,K\,h_2}{2}\,I_m^2 \right)\pi\,\sin 2\beta_o$$

and total developed torque due to 3 phases

$$T = -\frac{3}{2}(K\,h_2\,I_m^2)\sin 2\beta_o \qquad (AP7.16)$$

Substituting for K in eqn.(AP7.16)

$$T = -\frac{3}{2}h_2\,I_m^2\,\frac{\pi\,N_s^2\,L\,r\,3\mu_o}{8}\sin\,2\beta_o$$

or $\qquad\qquad T = -K\,h_2\,I_m^2\,\sin 2\beta_o \qquad\qquad (AP7.17)$

where the *machine constant* K is now given by

$$K = \frac{3^2\,\pi\,L\,r\,\mu_o\,N_s^2}{16} \qquad (AP7.18)$$

or, in general, for an m-phase winding

$$K = \frac{m^2\pi\,L\,r\,\mu_o\,N_s^2}{16} \qquad (AP7.19)$$

Hysteresis Torque

As noted, hysteresis torque in the machine is considered to be obtained independently based on the alternating hysteresis loss in the rotor ring, now assumed fitted round the shuttle-shaped reluctance rotor[1]. The field configuration of Fig.AP7.1 still applies as in a hysteresis motor with the stator mmf leading the rotor mmf or axis by a certain angle γ being the angle of hysteretic advance. If now the resultant rotor mmf, during independent operation as hysteresis motor alone, is assumed to be the same as the axis of the magnetic arbor, the angle β_o emerges with a special significance if related to γ.

It is known that the hysteresis torque depends on $\sin \gamma$ and larger the value of γ the greater will be the developed torque due to hysteresis effect. The usual range of γ for most permanent-magnet materials is from $10°$ to $50°$, corresponding to the condition of 'low' magnetisation to saturation, with a value of about $40°$ being common during 'normal'-excitation operation. This is close to the value of β_o (=45°) for maximum reluctance torque. It therefore implies that when the machine is designed to develop maximum hysteresis torque, the reluctance torque obtainable from it would also be nearly maximum, indicating the best possible design[2].

---

[1]Note that a circumferential-flux hysteresis machine is considered.

[2]Observe that, with reference to Fig.AP7.1, the total developed torque is in the clockwise direction whereas the applied load or torque acts in the counter-clockwise direction.

The torque expression

Consider the maximum field intensity in the airgap when only the reluctance rotor is present. First obtain the maximum value of B_θ using eqn.(AP7.8) and from trigonometric considerations. In the airgap, H_θ is then related to B_θ by μ_o. Thus, considering only the terms of fundamental frequency, from eqn.(AP7.8)

$$B_\theta = K_1\left[h_o\sin(\omega t - \theta) - \frac{h_2}{2}\sin(\omega t + 2\beta_o - \theta)\right] \qquad (AP7.20)$$

where
$$K_1 = \frac{3\mu_o\, N_s I_m}{4} \qquad (AP7.20a)$$

Eqn.(AP7.20) implies that B_θ will be maximum when
$$\omega t - \theta = \pi/2 \qquad (AP7.21a)$$
and
$$\omega t + 2\beta_o - \theta = 3\pi/2 \qquad (AP7.21b)$$

Using the value of $(\omega t - \theta)$ from eqn.(AP7.21a) in eqn.(AP7.21b) would give $\beta_o = \pi/2$ which is un-realisable being the unstable operating condition. Therefore substituting for $(\omega t - \theta)$ from eqn.(AP7.21a) in eqn.(AP7.20)

$$B_\theta = K_1\left[h_o - \frac{h_2}{2}\sin\,(\pi/2) + 2\beta_o\right]$$

or
$$B_{\theta(max)} = K_1\left[h_o - \frac{h_2}{2}\cos 2\beta_o\right] \qquad (AP7.22)$$

is the maximum value of B_θ in the airgap at any instant.

Note that when $\beta_o = \pi/4$, $B_{\theta(max)} = K_1 h_o$.

The maximum field intensity is then given by
$$H_{\theta(max)} = B_{\theta(max)}/\mu_o$$
and the alternating hysteresis loss in the rotor ring is assumed to correspond to $H_{\theta(max)}$.

Now let the hysteresis loss/unit volume/rev be W_h joule, obtained, for example, from the area of the hysteresis loop for the rotor ring material.

Also, let speed of the rotor be n_s rpm and volume of the hysteresis ring material be V_{ring}.

Then total hysteresis power loss
$$P = W_h\, n_s\, V_{ring}\ \text{W syn.}$$
and the *theoretically* developed torque due to hysteresis[3].

$$T_h = \frac{P}{2\,\pi\,n_s}\ \text{Nm}$$

$$= \frac{W_h\, V_{ring}}{2\,\pi}\ \text{Nm} \qquad (AP7.23)$$

and is independent of speed.

---

[3]See, for example,

D.G.Young: The hysteresis clutch, Jour. I E E, Vol. 9, 1963, pp 437-39.

8

Computation of developed torque for a two-pole hysteresis-reluctance motor

The following machine constants and parameters are chosen arbitrarily

Mean airgap radius	: 20 mm, 0.02 m
Axial length	: 40 mm, 0.04 m
Minimum mechanical airgap, δ	: 0.254 mm, 0.254×10^{-3} m (0.010")
Radial thickness of hysteresis ring, t	: 1.0 mm, 0.001 m
No. of stator turns/phase, N_s	: 350
Rms current/phase	: 0.5 A or $I_m = \sqrt{2} \times 0.5$ A

A. Reluctance Torque

From eqn.(AP7.18)

$$K = \frac{3^2 \pi L r \mu_o N_s^2}{16}$$

Substituting

$$K = \frac{3^2 \pi \ 4 \times 10^{-2} \times 10^{-2} 4\pi \times 10^{-7} (350)^2}{16}$$

$$= 18 \times \pi^2 \times 3.5^2 \times 10^{-7} = 2.174 \times 10^{-4}$$

For $g(\theta)_{min} = t + \delta = 1.254$ mm, $1/g(\theta)$ when expanded in Fourier series gives

$$h_o = 0.4612 \times 10^3 \ m^{-1} \quad \text{and} \quad h_2 = 0.4388 \times 10^3 \ m^{-1}$$

Let β_o be assumed to be $40°$ which is taken approximately equal to γ under the assumed operating conditions. The reluctance torque is then given by eqn.(AP7.17).

Or

$$T_{rel} = K \ h_2 \ I_m^2 \ \sin 2\beta_o$$

$$= 2.174 \times 10^{-4} \times 0.4383 \times 10^3 \times 2.05^2 \ \sin 80°$$

$$= 4.7 \times 10^{-2} \ Nm$$

B. Hysteresis Torque

The volume of the rotor ring is

$$V_{ring} = \pi (21^2 - 20^2) \times 40 \times 10^{-9} \ m^3$$
$$= 5.15 \times 10^{-6} \ m^3$$

Using eqns.(AP7.20a) and (AP7.22)

$$B_{\theta(max)} = \frac{3\mu_o N_s I_m}{4}\left[h_o - \frac{h_2}{2}\cos 2\beta_o\right]$$

and

$$H_{\theta(max)} = \frac{3 N_s I_m}{4}\left[h_o - \frac{h_2}{2}\cos 2\beta_o\right]$$

$$= \frac{3 \times 350 \times \sqrt{2} \times 0.5}{4}\left[0.4612 - 0.2194\cos 80^{\circ}\right] \times 10^3$$

$$= 0.785 \times 10^5 \ A/m$$

If vicalloy is used for the rotor ring, the alternating hysteresis loss when acted upon by $H_{\theta(max)}$ = 0.785×10^5 A/m   would be 1.1×10^5 $J/m^3/rev$ approximately[1] and the hysteresis torque using eqn.(AP7.23) is

$$T_{hys} = \frac{1.1 \times 10^5 \times 5.15 \times 10^{-6}}{2\pi} \ Nm$$

$$= 9.02 \times 10^{-2} \ Nm$$

Then, total developed torque in the machine

$$T = T_{rel} + T_{hys}$$
$$= (4.7 + 9.02) \times 10^{-2}$$
$$= 13.72 \times 10^{-2} \ Nm$$

the contribution due to reluctance torque being about 35%.

---

[1]The material being driven well into saturation.

References and Bibliography

[1]   E.Warburg: Magnetic investigation, ANN. PHYSIC 3. Vol.13, 1881, pp141-64.

[2]   J.A.Ewing: On effects of retentiveness in the magnetisation of iron and steel, Proc. Roy, Soc., Vol. 34, 1882, 83, pp 39-45.

[3]   H.A.Rowland: On magnetic permeability and the maximum of magnetism or iron, steel and nickel, Phil. Mag., Vol.46, 1873, pp 140-59.

[4]   J.A.Ewing: On the magnetic susceptibility and retentiveness of soft iron (and steel), Phil. Trans. Roy. Soc., Vol. 176, Pt. II, 1883, pp 523-640.

[5]   J.A.Ewing: Contribution to the molecular theory of induced magnetism, Proc. Roy. Soc., Vol. 48, 1890, pp 342-58.

[6]   J.W.Rayleigh (Lord): On the energy of magnetised iron, Phil. Mag., Vol. 22, Series 5, 1886, pp 175-83.

[7]   J.A.Fleming: Note on magnetic hysteresis, The Electrician, Sept. 14, 1888, p 586.

[8]   J.Swinburn: The electrical papers at the British Association, Industries, Vol. 9, 1890, p 289

[9]   F.G.Baily: On hysteresis in iron and steel in a rotating magnetic field, Brit. Assoc. Rep., 1894, pp 576-77.

[10] E.A.Nesbitt: Vicalloy – A workable alloy for permanent magnets, Trans. A I M M E, Vol. 166, 1946, pp 415-25.

[11] S.Miyairi and T.Kataoka: A basic equivalent circuit of the hysteresis motor, Elect. Eng. In Japan, Vol. 85, (10), 1965, pp 41-50.

[12] G.Wakui and Y. Kusakari: Synchronous pull-out torque of hysteresis motor, ibid., (6), 1965, pp 31-42.

[13] S.Miyairi and T.Kataoka: Analysis of hysteresis motors considering eddy current effects, ibid., Vol. 86, (6), 1966, pp 67-77.

[14] S.A.Stoma: Effect of a non-sinusoidal magnetic field in the gap on the starting torque of a hysteresis motor, Elect. Tech., U S S R, Vol. 4, 1968, pp 15-27.

[15] S.D.T.Robertson and S.Z.G.Zaky: Analysis of the hysteresis machine – Pt. I, I E EE Trans., Vol. PAS-88, 1969, pp 474-83.

[16] M.A.Copeland and G.R.Slemon: An analysis of the hysteresis motor: I-Analysis of the idealised machine, ibid., Vol. PAS-82, 1963, pp 34-42.

[17] M.A.Copeland and G.R.Slemon: ibid., Vol. PAS-83, 1964, pp 619-25.

[18] E.Olsen: Applied Magnetism: a Study in Quantities (book), Ch 7, Philips Technical Library, 1966, pp 54-55.

[19] G.Ferraris: Rotazionielectrodinamiche prodotte per mezzo di correnti alternate, Atti. Dello R. academia d scienze, Vol. 23, 1888, pp 360-75.

[20] G.Ferraris: On the difference of phases of currents, on the retardation of induction and on the dissipation of energy in transformers – experimental and theoretical researches, The Telegraphic Journal and Electrical Review, Vol. 22, Feb. 3, 1888, pp 111-13.

[21] C.P.Steinmetz: Theory and Calculation of Alternating Phenomena (book), 1897, W.C.Johnstone. (III Ed. 1900, pp 293-96).

[22] C.P.Steinmetz: Theory and Calculation of Electrical Apparatus (book), Ch. 10, McGraw Hill Book Co., 1917, pp 168-71.

[23] H.Zipp: Hysteresis loss in polyphase induction motors near the speed of synchronism, Elekt. und Masch., Vol. 26, 1908, pp 443-50.

[24] T.Lehmann: The sudden change of power of induction motors on passing through synchronism, E.T.Z., Vol. 31, 1910, pp 1249-53.

[25] G.Vallauri: Phenomena accompanying the passage of induction motors through synchronism, Elekt. und Masch., Vol. 30, 1912, pp 1061-67.

[26] C.F.Smith: The experimental determination of the losses in motors, Jour. I E E, Vol. 39, 1907, pp 437-83.

[27] D.Robertson: Rotor hysteresis in polyphase induction motors, Electrician, Vol. 68, 1911, pp 12-14.

[28] W.Holmes and E.Grundy: Small self-starting synchronous time motors, Jour. I E E, Vol. 77, 1935, pp 379-99.

[29] B.R.Teare: Theory of hysteresis motor torque, Trans. A I E E, Vol. 59, 1940, pp 907-12.

[30] 'S.Smith & Sons': British Patent: 576 249, 1944.

[31] H.C.Roters: The hysteresis motor – advances which permit economical fractional horse power ratings, Trans. A I E E, Vol. 66, 1947, pp 1419-30.

[32] A.N.Larionev et al: General problems of the hysteresis motor theory, Electrichestvo, No. 7, 1958, pp 1-6.

[33] S.Miyairi and T.Kataoka: Airgap flux density distribution in hysteresis motors, Elect. Eng. In Japan, Vol. 83, 1963, pp 1-10.

[34] M.A.Rahman et al: An analysis of the hysteresis motor: III – parasitic losses, IEEE Trans., Vol. PAS 88, 1969, pp 954-61.

[35] H.E.Jaeschke: The hysteresis motor, Elekt. und Masch., Vol. 60, 1942, pp 176-88.

[36] A.M.Langen: Theory of ideal hysteresis motors, Elect. Tech., U S S R, Vol. 2, 1969, pp 107-18.

[37] S.Nakamura et al: Design of hysteresis motors, Elect. Eng. In Japan, Vol. 88, (6), 1968, p 11-20.

[38] P.L. Alger and W.R.Oney: Torque-energy relation in induction machines, Trans. AIEE, Vol. 73, Pt. IIIA, 1954, pp 259-264.

[39] M.R.Harris and W.Z.Fam: Analysis and measurement of radial power flow in machine airgap, Proc. IEE, Vol. 113, (10), 1966, pp 1607-1615.

[40] E.I.Hawthorne: Flow of energy in direct current machines, Trans. AIEE, Vol. 72, Pt. I, 1953, pp 438-445.

[41] 'Crouzet Co. Ltd.', France: British Patent: 1 046 016, 1966.

[42] J.W.Rayleigh: The behaviour of iron and steel under the operation of feeble magnetic force, Phil. Mag., Vol. 23, Serues 5, 1887, pp 225-45.

[43] R.Hiecke: Hysteresis in a rotating field, Elwcrotechn. Zeitschr., Vol. 23, 1902, pp 142-43.

[44] F. Brailsford: Rotational hysteresis loss in electrical sheet steels, Jour. I E E, Vol. 83, 1938, pp 566-75.

[45] E.C.Stoner and E.P.Wohlfarth: Mechanism of magnetic hysteresis in heterogeneous alloys, Trans. Roy. Soc., Vol. 240, Series A, 1948, pp 599-644.

[46] J.H.Poynting: On the transfer of energy in the electromagnetic field, Phil. Trans., Roy. Soc., Vol. 175, Pt. II, 1885, pp 343-61.

[47] J.Slepian: Energy flow in electric systems – the V-I energy flow postulate, Trans. AIEE, Vol. 61, 1942, pp 835-41.

[48] R.W.Fountain and F.L.Joseph: Development of mechanical and magnetic hardness in 10 pct. V-Co-Fe alloy, Trans. A I M E, (Feb), 1953, pp 39-56.

[49] D.L.Martin and A.H.Geisler: Constitution and properties of cobalt-iron-vanadium alloys, Trans. A S M, Vol. 44, 1952, pp 461-83.

[50] W.J.Karplus: Analog Simulation (book), Ch. 5, McGraw Hill Book Co., 1958, pp 127-29.

[51] D.Vitkovitch: Field Analysis: Experimental and Computational Methods (book), Ch. 5, Van Nostrand Co., 1966, pp 199-203.

[52] M.Abramowitz and I.A.Stegun: Handbook of Mathematical Functions with Formulas, Graphs and Mathematical Tables, Dover Publications, 1965, pp 590-92.

[53] R.Bulirsch: Numerical calculation of elliptic integrals and elliptic functions, Numerische Mathematik, Vol. 7, 1965, pp 78-90.

Index

Author Index

Notes

Notes

Notes

For Product Safety Concerns and Information please contact our EU
representative GPSR@taylorandfrancis.com
Taylor & Francis Verlag GmbH, Kaufingerstraße 24, 80331 München, Germany

www.ingramcontent.com/pod-product-compliance
Lightning Source LLC
Chambersburg PA
CBHW081055220326
41598CB00038B/7103